SIR DAVID FROST is the only person to have interviewed the last seven presidents of the United States and the last six prime ministers of Great Britain. He has been awarded all the major television awards, including the Emmy Award (for *The David Frost Show*—twice) in the United States and the BAFTA Fellowship in the United Kingdom, their highest honor.

Often described as "a one-man conglomerate," Sir David has worked variously as an author, film and television producer, publisher, lecturer and impresario, and co-founded two network companies in the United Kingdom, LWT and TV-am.

Sir David hosts *Frost Over the World*, his weekly current affairs program for Al Jazeera English, as well as series for the BBC and ITV. He is the executive producer of a remake of the film *The Dam Busters*, with Peter Jackson and Universal.

He lives in London, Hampshire, and on British Airways.

FROST/NIXON

ALSO BY SIR DAVID FROST

That Was the Week That Was

How to Live Under Labour

Talking with Frost

The English

The Presidential Debate 1968

The Americans

Whitlam and Frost

I Gave Them a Sword

I Could Have Kicked Myself (David Frost's Book of the World's Decisions)

Who Wants to Be a Millionaire?

The Mid-Atlantic Companion (with Michael Shey)

The Rich Tide (with Michael Shey)

If You'll Believe That

The World's Shortest Books

Billy Graham: Thirty Years of Conversations with David Frost

David Frost: An Autobiography: Part One: From Congregations to Audiences

FROST/NIXON

BEHIND THE SCENES OF THE NIXON INTERVIEWS

SIR DAVID FROST

WITH BOB ZELNICK

NEW YORK • LONDON • TORONTO • SYDNEY

HARPER PERENNIAL

FIRST EDITION

Designed by Jaime Putorti

Library of Congress Cataloging-in-Publication Data
is available upon request.

ISBN: 978-0-06–144586-6
ISBN-10: 0-06-144586-X

08 09 10 11 DIX/RRD 10 9 8 7 6 5 4 3

CONTENTS

AUTHOR'S NOTE

Frost/Nixon is a sequel to my earlier book on the Nixon interviews, *I Gave Them a Sword,* written in 1977. Wherever relevant, this book draws upon the resources of that earlier book, but thirty years later, there is more to talk about—not least the story of Nixon in retirement and an assessment of the Nixon presidency as seen from the vantage point of 2007 rather than 1977.

I have also included, for the first time, five transcripts from discrete parts of the interviews—not only Watergate and Vietnam but also the Huston plan, Henry Kissinger, and Chile.

I hope you approve of the result.

INTRODUCTION

While Watergate is only a part of this book, it is a very important part. I thought therefore that, as an introduction to younger readers and as a reminder to older readers, it might be helpful to summarize the salient facts about it before we begin.

WATERGATE—WHY IT MATTERED

On the night of June 17, 1972, a security guard at the plush Watergate office and apartment complex alerted the Washington, D.C., police that burglars had entered the building and were apparently still on the premises. Responding quickly, the police encountered five men about the offices of the Democratic National Committee. They had come to repair a listening device placed weeks earlier on the phone of DNC National Chairman Lawrence O'Brien. A second "bug" had been installed on the phone of a senior campaign official, Spencer Oliver.

What the police did not know, and what it would take weeks of investigation for them to find out, was that the five burglars—four of them Cuban veterans of CIA operations—were being directed from a room at the Howard Johnson hotel across the street by a senior official of the Richard Nixon reelection campaign, G. Gordon Liddy, and by a White House consultant and career CIA veteran, E. Howard Hunt. Almost immediately the police and the FBI began trying to learn who had been involved in planning the burglary, while the White House began trying to prevent them from finding out. In short, the White House went into a cover-up mode with the clear intent of obstructing justice.

From the moment in March 1973 when one of the five initial Watergate burglars disclosed, in a letter to the presiding trial judge, that perjury had been committed to shield the criminal involvement of more senior officials of the Nixon administration, until August 9, 1974, when Nixon resigned the presidency in disgrace, Watergate became a U.S. cause célèbre, a national obsession, or perhaps both. Many friends of the United States around the world thought we had taken leave of our senses. How could we show such disdain for the man who had brought us "peace with honor" in Vietnam, détente and arms control agreements with the Soviet Union, and a big start toward normal relations with the People's Republic of China? What was Watergate other than the "third-rate burglary" contemptuously dismissed by Mr. Nixon's press aide? Didn't the country know that America's security was put at much greater risk by those who would cripple its leadership than by those engaged in a nasty but not altogether unprecedented political prank? It is fair to say that even today many Americans too young to have shared the experience of Watergate with their parents or grandparents might be posing the same questions.

There are several answers. First, there is nothing small or insignificant about a number of rather senior officials from the Nixon administration and campaign gathering in the office of the attorney general of the United States to discuss such political operations as wiretapping phones, planting office bugs, using prostitutes, and spreading outrageous lies. Such matters go to the integrity of the political process, no small matter in a democracy. Ringleader

Howard Hunt's White House safe, for example, contained forged cables designed to show that the martyred Democratic President John F. Kennedy knew in advance that those involved in a coup against South Vietnamese President Ngo Dinh Diem planned to kill him.

Second, in joining the cover-up early on, Nixon threatened to corrupt important agencies of government. Asking the CIA to pull the FBI off the investigation was no trivial exercise of politics as usual. It was instead the abuse of two of the nation's most secret and important security agencies, which cannot afford to squander the public trust with which they have been invested.

Third, the Watergate investigation brought to light other abuses of power threatening the rights of Americans to be secure in their homes and offices, rights guaranteed by the U.S. Constitution. For example, in approving the "Huston plan" for burglaries without court warrants against those suspected of plotting violent or other illegal activities, Nixon was usurping the critical historical role of the judicial branch of government. From there it was a small step to burgle the office of the psychiatrist treating Daniel Ellsberg, the man who leaked the so-called Pentagon Papers chronicling the decisions that led to the massive U.S. intervention in Vietnam. Medical and psychiatric records are, of course, privileged and cannot be introduced into the public record without the patient's consent.

Concerns generated by these presidential usurpations came dramatically home to U.S. citizens one October evening when Nixon ordered his attorney general to fire Special Prosecutor Archibald Cox when the latter refused to obey an order by the president to abandon his subpoena for Watergate conversations preserved by the White House taping system. The attorney general and his deputy both resigned rather than obey the presidential order. While the third in command executed the order, the national loss of support for Mr. Nixon had by then passed the point of no return, and he was forced by a Supreme Court decision to turn over the tapes. The tapes proved to be damning in both content and tone. Once they entered the public realm, few doubted that Nixon was through.

There is, of course, a lot more to be said about Watergate. Although coming off the heels of an unprecedented reelection victory, Nixon had long been known as a political "gut fighter," who sought no quarter and provided none to his foes. Even so, had he chosen early on to dismantle his taping system and destroy the incriminating documents, he could certainly have escaped the serious threat of impeachment. That he failed to do so was probably a function of both political arrogance and bad advice.

Watergate also represented the high point for the role of the media as watchdog over the country's democratic system, a role it had also played during the civil rights era and the disaster in Vietnam. The investigative reporters Robert Woodward and Carl Bernstein gained immortality in their profession for disclosing important details about the financial and political trails, details that kept the investigation alive when—but for their work—it would almost certainly have faltered.

Finally, Watergate taught Americans something about themselves, their respect for the rule of law, their willingness to rise above partisanship in common battle to defend the institutions of freedom. Looking back at Watergate a year after Nixon left office, one columnist gushingly recalled "our moment of shared wonder and love of country." That may be overstating it a bit. Over the centuries Americans have shown some dubious political traits as well as noble ones. They have elected a Buchanan for every Lincoln, a Johnson (take your pick) for every Roosevelt (take your pick there, too). Watergate suggests, however, that while the United States may not be immune to the wiles of the political huckster, it remains tough prey for the would-be tyrant.

Part I

FROST/NIXON

1

THE DEAL

"It will be a sort of intellectual *Rocky.*"

The speaker was the writer Peter Morgan, and the time was January 2004. Peter and his producer, Matthew Byam-Shaw, had come to my office to talk about their idea for a stage play, to be called *Frost/Nixon,* which would tell the story of the Nixon interviews and Nixon's dramatic mea culpa in 1977.

They had three main requests. First: As the holder of the rights, would I give them permission to go ahead with the project? After some discussion, I said that, in principle, I would. That led to the second request: Would I let them have these rights for nothing? Peter and Matthew are both charming and persuasive, as you can tell by the fact that I said yes to this request. *Frost/Nixon,* they hoped, would open at the Donmar Warehouse in London and hopefully transfer into the West End. I said that my free grant of rights would extend to both these eventualities but not to any further manifestations of *Frost/Nixon.*

Oddly enough, it was not the money, it was the third request that gave me the most pause. Peter said that they both thought that *Frost/Nixon* would have more credibility if I had no editorial control. That was more difficult, and I said that I needed time to think about it.

It was a couple of months later before I gave them the green light on this issue. I felt very fifty-fifty about it at the time because I would be entrusting a project that was very precious to me to third parties. On the other hand, they felt that the play would get a better hearing if it were independent of my or my company Paradine's editorial control. In the end, I decided that the advantages probably just about outweighed the disadvantages, though when I saw the first draft, I was not so sure. Later drafts upset me less. I think that was because they were an improvement, or maybe I was just getting inured to the experience!

It is a curious feeling to go to the theater and watch yourself onstage—particularly if the "Frost" character is depicting some of the most dramatic episodes of your life. They were events that had taken place thirty years before, but somehow it did not feel that way. Peter had promised that he could make these events seem relevant, even current, and he had achieved that.

I attended a preview of *Frost/Nixon* two or three nights before the play opened in August 2006. I thought it was brilliantly written, directed, and acted. There were more fictionalizations than I would have preferred, although one such piece of fictionalization—Nixon's phone call to me on the eve of Watergate—was, I thought, a masterpiece.

I was not so sure about some of the other fictionalizations. Why was Watergate now the twelfth of the twelve sessions and not—as actually happened—two sessions in the middle, at sessions eight and nine? Why did James Reston's discoveries from the Watergate tapes only reach me on the morning of the Watergate session and not eight months earlier, as had actually been the case? Why did the early sessions, which contained a lot of good material, have to be depicted so negatively? Why do we see Swifty Lazar, Nixon's agent, making a series of demands without learning that they had been successfully rejected? Whenever I made these points to Peter, he would simply sigh and say, "David, you've got to remember this is a play, not a documentary." However, aware of my concern, he thoughtfully added an author's note to the program, making the point that he had sometimes found it irresistible to let his imagination take over.

And the play was an instant hit. The rave reviews were unanimous and Peter, the director, Michael Grandage, and both Michael Sheen ("Frost") and Frank Langella ("Nixon") were deservedly saluted.

Frank Langella did not look like Nixon, but he *was* Nixon. "I have never been a Method actor," he told me. "Normally, when I'm offstage during a Broadway play, I chat to the stage manager about how the Mets are doing or whatever, but with this play, the tension is such that I did not want to go out of character, even for a minute, when I was offstage. I would go to the darkest corner at the back of the stage and just stay with my thoughts and wait. When I was required, the stage manager had to come over to me and say, 'Mr. President, you are needed onstage.' "

I met Michael Sheen for the first time after attending that preview. The cast had not been told that I was there. Michael said that they were all bewildered because for the first twenty minutes the audience seemed nervous and there was less response than usual. I don't know whether people expected me to leap up and say, "Stop! That's not true!"

When I interviewed Michael last December, shortly after the Broadway production and the film had been announced, Michael said, "I'm going to be playing David Frost for the next year."

"That's a coincidence," I said. "So am I."

How did the Nixon interviews come to pass in the first place? Well, I must say that as I look back now, I marvel at the fact that we managed to pull them off. There were so many obstacles and challenges to overcome.

First, there was the challenge of getting Richard Nixon to say yes.

"Don't waste your time," said an Australian, adding cheerfully, "you've got Buckley's"—a piece of Australian vernacular intended to make a lost cause seem roseate by comparison.

"In the words of David Schoenbrun, talking about a possible interview during World War Two," I replied, " 'let de Gaulle say no.' "

I knew from experience that getting a clear response—whether yes or no—would not be easy. Experience came from *The David Frost Show.* Following the interview with then candidate Nixon that I had done in 1968, we would make annual requests for the president to appear on the program. The annual White House response had an almost ritual quality to it. It would be signed by Mr. Nixon's press secretary, Ron Ziegler. Always Ziegler would begin by saying, "I accept your invitation for the President to appear on a show with you." And, always, after "accepting" the invitation, Ziegler would state that the question of if and when to actually *make* the appearance on the show would be taken up with the president, with further information to be provided should Mr. Nixon actually agree to be interviewed.

This touching little habit of accepting pieces of paper on which invitations were written without responding affirmatively to the invitations themselves, I came to accept as a wholly innocent indication of Ziegler's ability to render the English language inoperative, even in matters not involving alleged presidential culpability. Just once, though, I would have liked to have Ziegler reject my invitation and Mr. Nixon agree to be interviewed.

But that, of course, never came to pass. Neither, understandably enough, was there an immediate response to my phone calls, now to San Clemente, California. The breakthrough came in late June 1975, when the *New York* magazine publisher, Clay Felker, returning from a weekend in the Hamptons, telephoned to say that he had encountered Swifty Lazar at a party. Swifty Lazar was the representative that Richard Nixon was said to be using. Clay said that he had gained a distinct impression that Swifty was now authorized to act for Nixon in the area of television as well as that of books, and that, indeed, one of Swifty's purposes in visiting the East Coast was to see what sort of interest in a Nixon interview he could whip up among the three networks. I knew I had to move quickly—and alone. Apart from Felker and a few close colleagues, the Nixon interviews were something I had not spoken to anybody about—partially for fear of being declared certifiable but more because I didn't want to give the idea to anyone who did not have it already.

I was glad that I was dealing with Swifty Lazar. Noted for his legendary ability to enter a revolving door behind you and come out in front, Swifty believed in getting right to the point. He wanted $750,000 for his client for a maximum of four one-hour shows. The main competitor—later revealed to be NBC—was currently on $300,000 and on its way to $400,000 for two hours and would not guarantee more than two hours. That seemed to me to be a heavy rate per hour—and an underestimate of how much Nixon had to offer, in terms of both information and public interest. I said I was thinking of a maximum of $500,000 for a minimum of four hours. Before returning to the question of a fee, however, I ticked off the points I regarded as mandatory.

First, the point on which I expected the most trouble: editorial control. I must have sole control of editing and content; Nixon must have none. Given the history of Swifty's client's relations with the media, it was a tall order, I knew, but it was essential. On the question of editorial control, Swifty would check with his client.

Second: Watergate. I knew that the recently announced Warner book deal contained no specific reference to Watergate at all. Reports on other approaches suggested that the Watergate factor might have been a problem in those negotiations. But regardless of all that, I had to have a cast-iron assurance that Watergate would be the subject of one of the four shows. That was new, and Swifty would investigate.

Third: exclusivity—before and after—was a must. An independent venture ran far greater risk, and we could just not afford to take, say, a $2 million risk and then be undercut at the last moment by some network with a valid-sounding interview pretext.

Fourth: time for interviewing. Although Swifty was talking about four hours on the air, I would want the right to many more hours of taping than that, a ratio of at least four to one, in case Nixon should ramble, or stern and persistent cross-examination proved necessary. "Sixteen hours," Swifty mused. He could see the point, but that was asking a heck of a lot.

Finally, I told Swifty that there was one other vital point. The book for which he had negotiated such a large contract—when was it due?

"Delivery of the whole book," Swifty told me, "or one of two books, is due in October 1976 with publication the following year."

"Well," I said, "we have to ensure that the television interviews precede the book—and the serialization of the book—by a minimum of three months."

"Are you sure?" asked Swifty. "That might cause me problems with Warner Bros., and after all, they have a first option on the television rights too."

"Yes, but they must have passed on those, Swifty."

"True—but they might reconsider."

"Well, that is their right." As far as I was concerned, the television had to come first.

"A lot of these points are new," said Swifty. "I'll be back to you." I gave him my phone numbers in London and waited.

Within days, the word came back: the response was not unfavorable. Swifty, God bless him, felt "duty-bound" to tell me that the "rival quote" was now $400,000 for two hours and then returned to his magic figure of $750,000. I said I could not really go beyond my original figure unless I had more time on the air. We compromised at $600,000 plus 20 percent of the profits, if any, for four ninety-minute shows (rather than one-hour shows), with $200,000 of that to be paid on signature.

However, the financial side of our negotiations took the smallest amount of time. Now we had to turn to the other points, almost any of which could break the deal.

First the sine qua non: What was the position on absolute editorial control? I waited for an explosion.

"Agreed," barked Swifty.

"He does realize that means he has no right to know the questions in advance?" I asked somewhat incredulously.

"Of course," said Swifty, "but I think he also realizes that the bona fides of these interviews have to be demonstrable if they are to have any impact at all."

The former president was worried about the exclusivity.

"Other television and radio interviews being out is understandable, but how about one- or two-minute statements for the news bulletins?" asked Swifty.

I took a deep breath. I rarely seemed to have the time to look back to my childhood, but for a moment I wanted to pinch myself that a Methodist minister's son from Beccles, Suffolk, was really laying down conditions like this for the former president of the Western world. I confirmed to myself that indeed I was.

"Only by mutual agreement," I told Swifty.

Next, did the former president understand the need for me to be protected from the book?

"Yes, he does," said Swifty, to my relief. Though, naturally, the former president felt very strongly that the publishers had to be protected too when it came to the Watergate period. "Watergate," said Swifty, "is the main problem."

And it had all been going so well, I thought. But then Swifty amplified his point. It was not that the former president had any objection to Watergate being part of the contract, it was just that he could not possibly speak out freely on the subject as long as he might affect the appeals of John Mitchell, H. R. Haldeman, and John Ehrlichman, which were still in progress. It seemed a fair point and we wrestled with the principles over the telephone, reaching a broad agreement, which was eventually enshrined in a cautious and complex clause.

August 9 was the chosen date for the parties to meet at San Clemente to sign the contract. Coincidentally, it was the first anniversary of Nixon's resignation. Nixon, dressed in his familiar dark blue suit, was waiting for us in his office. His handshake was firm, his gaze steady, his voice relaxed and confident. He had gained weight. He looked good—reassuringly good—to someone who was about to have to start a worldwide search for life insurance.

We exchanged pleasantries. Small talk. Always the most difficult part of any conversation with Richard Nixon. But today there was news in the papers of Leonid Brezhnev. Clutching at straws, I mentioned it.

"I would not like to be a Russian leader," said Nixon, shaking his head. "They never know when they're being taped."

Not a hint of a smile. Was he aware of the irony? Or just keeping the straightest face in the business?

"Communism stifles art," he said a moment or two later. "There is little important art you can cite from Communist countries. Solzhenitsyn is not nearly as impressive as Tolstoy."

But the purpose of the meeting was business. And for close to six hours, interrupted only by crab salad, Nixon paid attention to the task at hand. Finally, he asked his secretary to call in a tax attorney who had apparently been waiting in the wings to review the final document. As he entered, Nixon half smiled. "If I'd used this man four years ago, I wouldn't have gotten into all that trouble with the IRS!"

The moment came for signatures. And then the check. With a firm hand but a slightly trembling mind, I wrote the name "Richard M. Nixon" and then the words "Two Hundred Thousand Dollars" and then the numbers "$200,000."

Nixon reached for his billfold, but Swifty cut him short.

"Can I have the check, please?" he demanded.

"It's made out to me," the former president protested. "I'll deposit it."

"No, no, give it to me. That is the customary procedure."

"But what about the bank?"

"I'll take care of it."

"But, but—"

"Will you give it to me . . . *please,*" said Swifty, this time enunciating every word separately and distinctly.

Nixon handed the check to Swifty with the forlorn look of a little boy not allowed to consume the cookie he has swiped from the jar before dinner.

In the months that followed, Swifty had a falling-out with Nixon. Apparently he had been asked at a Hollywood party, "How on Earth can you represent a man like Richard Nixon?" And he had replied, "Listen, I'd represent Adolf Hitler if there was money in it."

Needless to say, the former president had not appreciated the linkage when it was related to him, and relations cooled.

It was a year or two later that I asked Swifty why on earth he had said that he'd represent Adolf Hitler if there was money in it. And he replied, "No, I never said that. I only said, 'I am not a literary censor, I would represent *Mein Kampf* if requested to do so.' "

I think I know which of the two quotes sounds more like Swifty to me, but it was probably a blessing for the project because Swifty loved his Hollywood parties and if you are an agent and you're behind the scenes, it's only by revealing what goes on behind the scenes that you can be the center of attention, and Swifty was always determined to be at least one of the centers of attention.

Had Swifty been present for the interviews, the two-month gap between the tapings and the transmissions would have presented him with an almost irresistible temptation.

The second major obstacle we encountered was the Nixon team. Later on, we grew to like the Nixon team, and our relationships worked pretty well. But we had certainly not reached that stage during the buildup to the interviews. A couple of examples . . .

When I met with Jack Brennan and Frank Gannon in Los Angeles, there was bad news. Richard Nixon had apparently fallen far behind his October 1976 deadline for his memoirs. Indeed, it now appeared that the memoirs would not be finished until April or May 1977. Breaking off for the months it would take him to prepare for the interviews, not to mention the months of arduous taping sessions, was totally unacceptable. There was just no way we could get to that business until May or June of the following year.

June 1977. That, I thought, would be a disaster. Agreements and contracts were due to be drawn up with the major networks in Great Britain, France, Italy, and Australia. This sort of unexpected delay would mean we would have to wait until the following season, and Brennan and Gannon must know that. What the hell did they think they were playing at?

"I'm afraid that's impossible," I snapped. "I have made commitments on the basis of your commitments to me. Even on our current schedule, one of my investors will have had his money tied up for eighteen months. That sort of delay is out of the question."

Brennan then unveiled his doomsday weapon: "If that's the case, then the president would prefer to return your check and call the whole thing off."

"Fine, Jack," I said, as calmly as I could. "I would say that will cost him between fifteen and twenty million dollars in damages growing out of his breach of contract. There are worldwide rights at stake now, not to mention our whole credibility. And I know what our rights are and what his obligation is. When we drafted the contract, we didn't leave ourselves vulnerable to this sort of game playing."

Brennan had no immediate answer but said he would carry my response back to Mr. Nixon and await further instructions. We parted company, and I flew to New York.

When Brennan called back, he said, "We've had discussions with the president and he is very anxious that you don't get screwed. He wants you to know that."

That was nice.

"And you were quite correct about the contract. It is your right to start taping after the election. That's fine, but obviously, also under contract, we won't be able to discuss Watergate yet. The Court of Appeals has yet to rule."

"Well, that's a great relief," I said. "I look forward to going in November and December with everything except Watergate if the Court of Appeals has not yet ruled."

"Yes," said Jack, "everything except Watergate. Which, as we define it, covers the break-in and the cover-up and also the resignation and the pardon and the final days. Because obviously they all bring up the whole question of guilt, which you can't discuss without discussing Watergate."

Before I could express my vehement disagreement with that interpretation, Gannon chimed in. "It also rules out the mind-set leading to Watergate, of course," he added. "All the security leaks, whether of a national security or a political security nature."

Needless to say, I could scarcely believe what I was hearing, but I might as well hear it all. What exactly did they feel the "mind-set" would exclude?

"Oh, the installation of the taping system, the Pentagon Papers, the early wiretaps, and the plumbers and Ellsberg of course. And anything from June '72 is difficult."

I am not quite sure how I managed to end the call in a civil fashion but I murmured something to the effect that we would have to discuss all this further.

Whichever way you looked at it, this was a pure wrecking operation. No Pentagon Papers. No Ellsberg. No plumbers. No wiretaps. No taping system. No mind-set leading to Watergate. What did they expect me to talk about for eighteen hours? The Postal Reform Act and the 1969 Ohio State–Purdue football game? We could be staring defeat in the face.

While I was abroad in July taping another series in Iran, John Birt, controller of current affairs for LWT, made a vital trip to San Clemente and made significant progress. I followed on September 9 with my colleague Marv Minoff to meet with Nixon. I explained to the president that I had to be back at Los Angeles airport to catch the 1 P.M. flight to Chicago for a lecture at the University of Northern Illinois.

"Are you getting paid?"

"Certainly."

"Just make sure you pay your taxes," he warned. "Otherwise you can get yourself in a lot of trouble."

Thanking him for the advice, I turned to the subject at hand. The prospect of waiting until May or June was devastating, I argued. That would mean we couldn't complete the editing until July and it would be August at the earliest until the shows would be aired. And no stations or advertisers would be confident of getting a large audience in August.

"I don't know about that," said Nixon. "We got a hell of an audience on August 9, 1974."

"Yes," I said, "but what do you do for an encore?"

We had another meeting at San Clemente on September 14; then a new clause of the contract was drawn up and mailed to San

Clemente on September 30. Then there were objections to that from the Nixon team and a new letter was sent off on November 3, but even that did not bring an immediate signature. So when I was in Los Angeles on December 7, I arranged to meet Frank Gannon for a drink and some hard discussion later.

I began by reviewing point by point the terms that had been derived from both the September 9 meeting, at which he'd been present, and the September 14 meeting, which he had missed. Clearly I was reciting nothing he hadn't seen with his own eyes or heard about at the time.

"Frank," I said, "the thing that puzzles me is that when we have these problems, always in the end the president acts honorably and helpfully. But for the life of me I can't understand why it's done in such a way as to invariably deprive him of any credit he might otherwise earn by being cooperative. I don't know if it's the advice he's given or what, but why are we getting f—ed around like this?"

Gannon said he would look into it. The next day we received word that the letter of understanding had been signed in the form submitted.

At last I had my contract, barely more than three months before the first interview session.

Third, we needed to assemble the team.

This was, of course, a category that was not so much an obstacle as a challenge.

The first priority was to locate a producer—or "coproducer," as we on this project would later call ourselves. The job definition, as I ticked it off, was daunting. My producer would not only have to be a first-rate journalist in his own right, to be able to command the respect of other first-rate journalists on the project, he had to be someone who could deal diplomatically with the Nixon people, who could make wise decisions fast under what might become incredible pressure, and who would constantly test my own instincts and conclusions. He had to be a conceptual thinker and, at the same time, know television technique and equipment as if it were second nature to him.

Did such a paragon exist? Fortunately I knew that at least one did. John Birt was the most outstanding current affairs producer I had ever worked with. He was now the controller of current affairs for LWT and, after weeks of discussion, he obtained a three-month leave of absence from November to January (which later had to be adjusted) to devote himself to the project. The quality of the rest of the team could also make or break the project. In June, I contacted the columnist Joseph Kraft, a longtime friend whose journalistic stature was attested to by the fact that he had been on more presidential hit lists than any other columnist in living memory.

Kraft first recommended James Reston, Jr., who had not followed his distinguished father onto *The New York Times* but was pursuing a successful career as a novelist and English instructor at the University of North Carolina. A bit later, Kraft recommended Robert Zelnick, a veteran reporter and, until recently, National Public Radio bureau chief, hardly known outside Washington but well respected by his colleagues. Bob would recruit a third reporter when we needed one and would generally act as the bureau chief of the smallest bureau in Washington.

After due consideration, he chose an outstanding investigative journalist, Phil Stanford, who had worked for us on abuses of power (non-Watergate).

We set July 1 as the starting date for Bob and Jim. John Birt would fly from London to Washington to meet them a few days later on July 12. The tempo of events was quickening. During May and June, the BBC in London had said yes, and Pacific Video of Los Angeles had agreed to become the facilities and technical unit for the production, deferring its fee of $290,000 to be recouped out of income.

During June, we had also found our "network erector." Syndicast Services would organize our network for us, deferring its fee of $175,000 in a similar way as Pacific Video. We set July 12 as the day on which we would announce that the special network was about to come into being. I was confident, but I could not help recalling the recent words of one reluctant noninvestor.

"The networks have said no," he told me. "The stations won't dare go it alone."

In Washington, Birt met first with Zelnick. From their initial handshake, they hit it off. Birt told Zelnick we wanted four "program briefing books," each dealing with a separate aspect of the Nixon administration. While the shape of the programs would be defined finally by the material generated by the interview sessions themselves, it was not too early to be thinking in terms of Nixon's foreign and domestic policies, Watergate, and the other abuse areas. There was also Nixon, the man himself. John asked Zelnick to prepare briefing books on Nixon's foreign and domestic policies.

Birt's first session with Reston, he found less encouraging. Jim regarded Nixon as the epitome of evil and, despite his resignation, a continuing threat to the American body politic. He felt that Nixon would know more about all areas of his presidency—including Watergate—than we could ever learn and thought it inevitable that he would win any confrontation between us based on an evidentiary interrogation. Jim was speaking partly as a novelist. What he seemed to want was a psychohistory of the Nixon presidency that would explain both the mind of Richard Nixon and the dark forces in American society that had carried him to the pinnacle of political power. Birt and Zelnick felt as I did: that it was the whole record of Richard Nixon that had to be examined. "The biggest danger," said Bob, "is in failing to do a thorough job and not putting this man on record on things about his presidency which he has never had to address in a comprehensive way."

I arrived in Washington on the fifteenth to find the argument still going on and Birt by now harboring serious doubts about whether he and Jim could ever work fruitfully together. I was all for providing insight into the Nixon character, if at all possible. But to make that the be-all and end-all was setting the sights far too low. However, I decided that Jim should stay on the team. If he lacked Birt's combination of intellect and television professionalism or Zelnick's tactical and logical intuition, he added a dimension of passion and creativity that could prove exceedingly valuable. It was a decision I would never regret. While Reston would continue to press doggedly with his lonely plea for a quasi-psychiatric interrogation of Nixon, he did a masterful researching and organizing of the Watergate material.

Indeed, by mid-September, while perusing the Watergate trial transcript at the Federal Courthouse on John Marshall Place, Jim came up with a gem. Leafing through the prosecution exhibits, he came upon a conversation between Nixon and Charles Colson dated June 20, 1972, just three days after the break-in. And other conversations on February 13, 1973, with Nixon saying, "When I'm speaking about Watergate, that's the whole point of the election: this tremendous investigation rests unless one of the seven begins to talk: that's the problem." And on February 14, Nixon says, "That's where we've got to cut our losses. My losses are to be cut. The president's loss has got to be cut on the cover-up deal."

These tapes were new, and the dates on which they had been recorded were devastating. It was the best kind of scoop—developed not through a leak or a breach of ethics on anyone's part but through Jim's own sheer shoe-leather diligence.

Fourth, we had to raise the money.

Finance—or the lack of it—was a perpetual subtext of the Nixon interviews. For instance, the bankers who had promised that first $200,000 of the Nixon check that we had to present on signature had reneged with just twenty-four hours to go. That crisis was averted when James Wolfensohn—then at J. Henry Schroder in

New York—came to the rescue. It was soon clear that the overall cost of the Nixon interviews was going to be in the region of $2 million and that was the sum we had to raise. Polygram, the German-Dutch musical entertainment conglomerate, soon came into the picture with the $200,000 that enabled me to repay Jim Wolfensohn. Later on it also came in for an additional $400,000.

Financier James Goldsmith said that he would match that sum. But it was soon clear that raising $2 million by financial contributions alone was going to be difficult, if not impossible. We had to have presales as well, around the world. I was in Sydney, Australia, when Kerry Packer of the National 9 Network said yes at dinner on the night of July 1. We would settle the details over the phone before I left for Los Angeles the next day. After completing a week's stint from 9 A.M. til noon on radio on Sydney's 2SM network throughout Australia, I called Kerry in his office. Kerry confirmed that he wanted to take the Nixon interviews but said he would not go a cent above $160,000. I said I would not come down a cent below $175,000 because at that stage every dollar mattered. An impasse like that—particularly between friends—can easily end in disaster.

"Let's toss for it," I said.

"Okay, Frostie," said Kerry. "You call . . ."

"Tails."

"Tails it is, old son."

I had equally fruitful conversations with TF1 in France, RAI in Italy, and the BBC in London. The gap was beginning to close, but then one other investor who had pledged his support withdrew and the gap began to open again.

Fifth, where would we broadcast the interviews?

Despite perfectly amiable conversations, it became clear that the three network news departments would not change their minds about broadcasting the Nixon interviews. NBC News, CBS News, and ABC News rarely, if ever, took independent productions from outside companies. In this case, with something that they could see would have considerable impact, they were even more adamant that they did not want to take on this independent production.

So I had to erect my own network, making syndication deals in each market in the country. It had never been done before, but it was the only way. I was fortunate enough to find a syndication distribution company, Syndicast, which was prepared to take on the challenge. Friday, July 16, 1976, was a crucial day. I met with Syndicast at 8:15 A.M. to check on the progress to date. The people there told me that in the previous four days, since July 12, two of the nation's most respected station groups—Scripps Howard, whose markets were Cleveland, Cincinnati, Memphis, and West Palm Beach, and Corinthian, with stations in Houston, Indianapolis, Sacramento, Fort Wayne, and Tulsa—had committed to the Nixon interviews. At 8:45 A.M., Syndicast had arranged a meeting with George Moynihan and Pat Polillo of Group W Westinghouse Broadcasting, who had produced *The David Frost Show* and whose markets were Boston, Philadelphia, Pittsburgh, San Francisco, and Baltimore. And at ten o'clock, with another old friend, Larry Fraiberg of Channel 5 in New York, who was representing the Metromedia group of stations in New York, Los Angeles, Washington, D.C., Kansas City, and Minneapolis. Both meetings went well, and by lunchtime both station groups had said yes. We had 40 percent of the nation in one week! That left only another 170 stations to go . . .

Sixth, there was the little matter of sponsors.

In order to construct a one-off network like this, the arrangement with the individual stations had to be as follows:

In the course of each ninety-minute program there would be twelve minutes of commercials. The local station would broadcast six minutes of our national commercials and in return have six minutes that they could sell themselves. So we had six minutes in each show to fill with our commercials.

As an adman explained to me, "Look, you've got a big problem. Out of the companies you're approaching, fifty percent wouldn't have wanted to be seen having anything to do with Nixon even when he was president, and the other fifty percent are trying to make people forget that they did."

Michael Johnson, the president of Syndicast, wrote to thirty major companies to see if they would be interested in sole sponsorship of all four programs. "The opportunity exists . . . to perform an unprecedented national service and to reach, at the same time, a very large and intense audience."

If the matter had not been so serious, the responses would have been hilarious. The thirty companies were unanimous. "We would have no interest whatsoever in programs of this type," said one. "I'm sorry to tell you that our television strategy over the next several years does not include these kinds of programs."

One advertiser's letter seemed to have two distinct authors: "While we agree with you about the outstanding historic nature of these broadcasts, I'm sure you can see it would be dangerous for us to be associated with an enterprise like that."

No sole sponsors, then.

It was Datsun who made the first verbal commitment for four thirty-second spots. The first signed agreement came on January 7 from Weed Eater in Houston. It obtained its reward on *60 Minutes* four months later, when it became a subject of national debate on that outstanding newsmagazine.

"Weed Eater," said Mike Wallace. "I don't know what Weed Eater is, but they have bought one spot, two spots?" I was not, I must confess, totally equipped to answer him.

> FROST: Weed Eater is a product that you're going to come to know and love and understand, but first I hope that I come to understand it. Let's be clear about this. We're seeking advertisers who realize it is history, but it's controversial

history. So we are seeking advertisers with courage, and these people have courage. But we are—

WALLACE: Weed Eater has courage?

FROST: Weed Eater has courage.

So did Richard Gelb of Bristol-Myers, who came in with a really positive contribution, but when we started taping the interviews in March, we still did not have a full lineup. Halfway through the taping of the interviews, we had the reassuring news that we were sold out.

And so all the obstacles were eventually—and sometimes at the last minute—overcome. Of course, while all of this business planning was going on, so was the research.

Bob Zelnick, Jim Reston, and Phil Stanford, our third journalistic contributor, were producing the briefing books in Washington. John Birt and I were working on those and of course on the tapes in London.

We were amateurs in the cloak-and-dagger business, but we did our best. We had strict confidentiality clauses for everyone in sight. We knew the name "Nixon" immediately attracted attention, so in telexes, we referred to "the subject," and when we met in restaurants, whenever waiters or others approached, Richard Nixon would become "William Holden" or "Charlton Heston," but I would be hard pressed to claim that we ever fooled anyone with our little moments of melodrama.

Leaving the table briefly during one of our sessions at the Rive Gauche in Washington, Zelnick encountered one of our regular waiters.

"Mr. Zelnick," he said, "I hate to interfere, but I have waited on

William Holden dozens of times, and he just doesn't seem like the sort of fellow who would tell witnesses to lie to juries."

Before Zelnick could think of a response, the waiter went on, "But Richard Nixon does, and I hope Mr. Frost gives him hell."

The Rive Gauche was about to give way to the delights of the Beverly Hilton. Unfortunately, Phil Stanford would not join us because of prior commitments, but otherwise we were at full strength. John Birt had arrived from London, followed a day or two later by Libby Reeves-Purdie, my London secretary and girl Friday. Marv Minoff, Bob Zelnick, and Jim Reston were commuting between the hotel and our offices in Century City, where Don and Susan Clark and Stewart Hillner were handling the awesome logistics associated with the project, poring over airline schedules, and trying to work out in advance how to dispatch tapes to Denmark to meet a deadline and get them transferred from 525 lines to 625 lines on the way. Over at Don Stern Productions, Jørn Winther, our director; Don Stern himself, who was to be our editor; and Tony Hudz had been reviewing thousands of feet of newsreel film and tape supplied by our film researcher in New York, Ann Dean. John had brought Bernard Lodge's titles and Dudley Simpson's music over from London, and he and Jørn were busy checking out the final technical arrangements with Pacific Video. It was a team, small and ad hoc though it may have been, good enough to fill any executive producer with confidence. And, as I write their names again now, gratitude.

After we had arrived in California, Jack Brennan and I had one more rousing confrontation before the interviews began. We were lunching at the Quiet Cannon near San Clemente. "How do we know," asked Brennan, "that you're not going to screw us with the editing?"

I demurred and quickly put forth the question that was on my mind: "How do we know you're not going to screw us with the stonewalling?"

We had both stated as boldly as possible our basic fears. We had not put each other's minds at rest, but at least it did make dialogue easier.

Brennan went on. "You know, sixty percent of what this guy did in office was right," he said, "and thirty percent might have been wrong but he thought it was right at the time."

I stared at Mr. Brennan without having to say a word. Both of us had passed arithmetic in elementary school. Ten percent of what Nixon had done was wrong and he knew it was wrong.

Brennan finally broke the silence. "If you screw us on the sixty percent, I'm going to ruin you if it takes the rest of my life."

I did not want to quibble over the exact percentages. Putting my arm around him, I replied, "and if you stonewall us on the ten percent, I'm going to ruin *you* if it takes the rest of my life."

It was a curious compact born of belligerence, but I found it oddly encouraging.

On the drive back from the Quiet Cannon to the Beverly Hilton, I was horrified to realize that I'd left my briefcase behind at San Clemente, containing at least one of our four basic briefing books.

I hastened to find Zelnick, only to discover that he'd already learned about the missing item from Ken Khachigian, a senior Nixon aide. "Ken just called to extend his thanks for your leaving the briefcase there," he told me. "He says it saved them a firebombing."

We continued talking to anybody and everybody who could help us understand Nixon better. A senior domestic policy adviser theorized to Bob Zelnick that there were really two Nixons: the one who was fascinated with great international issues and the mechanics of governing and the frighteningly insecure political thug. Each Nixon had surrounded himself with a predictable set of colleagues. In the end, the thugs had prevailed and been responsible for bringing down the Nixon presidency. But both Nixons existed. Like the hero in Hermann Hesse's *Steppenwolf,* Nixon could well complain, "Alas, two souls beat within my breast." Or, as Dr. Johnson said of one of his contemporaries, "He may do very well, as long as he can outrun his character: but the moment his character catches up with him, he's all over."

How many Nixons, I wondered, would we encounter in the coming month?

2

THE INTERVIEWS

THE INTERVIEWS: DAYS 1–2

March 23, 1977, and March 25, 1977

As N-Day dawned in California, I found myself considering how best to get the ex-president talking. Though the sessions were bound to be edited and aired out of order, I liked the idea of starting the first day with a question that might also head our first program. I also felt it was important for Nixon and the American people to pick up where they had left off three years earlier. As I drank a black coffee, my mind wandered toward Watergate.

I met my team in the hotel lobby at 7:45, and we set out for the one-and-a-half-hour drive to Monarch Bay. As we drove, we debated the merits of various possible openings.

"Why don't we start with the question everybody talks about?" I asked. "Why didn't he burn the tapes?"

Bob didn't like the idea. "It could open up the entire Watergate matter long before we're ready to delve into it. And if he regards it as a breach of our understanding to discuss Watergate after Easter, it's more likely to set the sort of negative tone you're worried about."

Birt was less apprehensive.

"Apart from Bob's point about the breach of understanding, I can't see any real harm in it," he said. "It's not likely to take him by surprise. My goodness, if he's thought of nothing else about Watergate, he's surely thought of that. Provided we don't stay too long on the subject, I don't feel strongly one way or the other." We made no final decision in the car. I would check with Brennan about the breach-of-understanding issue, then trust my own instincts.

Our 9:35 arrival at the Monarch Bay estate was greeted by a swarm of network and local media. I paused briefly, then hurried inside. Nixon would be arriving at 10:10.

At precisely the appointed moment, Nixon's white Lincoln Continental pulled up to the curb. What impressed me first about the ex-president that first morning was his apparent robust health. Gone was the tortured mien of the resignation period, the jangle of bones that had offered a hollow victory salute on his departure by helicopter, even the recuperating recluse of our meetings with Swifty Lazar. The Nixon now gliding toward us looked formidable, tough, knowledgeable, well briefed, and confident he would soon be at the top of his game. It was a moment to savor.

We shook hands, then parted for a few moments of final preparation and makeup. Fifteen minutes later, we walked along the passage and through the kitchen into the living room, where the cameras were waiting. Nixon showed no sign of nervousness. His face seemed set, determined; his gaze firm.

The first question could be so important, I thought again. Will the nudge of the unexpected elicit an honest response, crisply delivered, right to the point? Or will he feel betrayed, besieged by those hostile media types, the target of another gunman from the eastern establishment? As we took our seats, I decided to test him. The cameras rolled.

"Mr. President, we are going to be covering a lot of subjects and a great deal of detail over the course of the next six hours, but I must begin completely out of context by asking you one question, more than any other, almost every American and people all over the world want me to ask. They all have their questions, but one of them in every case is: Why didn't you burn the tapes?"

Nixon didn't like what he had heard. Intellectually, he was an orderly man who stored his information in modular units and who drew upon them as programmed. This day was supposed to be about other topics, and his interlocutor had started it off with a subject about which he knew everything but had no structured thoughts prepared. Still, Nixon wanted his public back and the seasoned politician knew the hazards of a brash response. The air of feigned diplomacy he now adopted was a hat he would don repeatedly over the next month when confronted by questions or accusations he found distasteful.

"Mr. Frost," he began now, "as you know, we agreed that we would cover the Watergate aspects of these various programs, the White House years, and the early life as well in our last taping session, but since you have that as a major concern among your listeners and viewers, there is no reason why I at least can't respond briefly to it now and you can explore it at greater length later if you like."

I noted the breach of protocol but returned to the question. "But everybody says, Why didn't you burn the tapes? And in a word, I wondered what your answer was."

A generation later, journalists, scholars, and others still ponder the question. Dates and events are tossed around like football plays on Super Bowl Sunday. All are premised on Nixon explaining his action as a defense of the presidency from encroachment by Congress and the courts. The earlier Nixon had acted, the less evidence documenting the Watergate crimes would have come to light and the more time his natural political allies would have had to rally to his defense. The optimum moment to move would have been within hours of White House aide Alexander Butterfield's

Ervin Committee testimony disclosing the existence of the taping system. Destruction of the tapes could have been accompanied by a speedy but thorough White House investigation identifying the wrongdoers—nearly all of whom were already known by Mr. Nixon—firing them from whatever positions they held, and issuing pardons to all involved in a "boys will be boys" type adventure that had gotten horrifyingly out of hand.

Back on the set, I hoped silently that Mr. Nixon could be both forthright and concise in his response. It was not to be. Nixon warned that "it takes a little bit more than a word" because "you have to understand why they were put in." Twenty minutes later we were still understanding a pretty simple narrative. Upon moving into the White House, Nixon had learned that Lyndon Johnson had maintained an elaborate taping system including both phones and offices, apparently his way of building an accurate record of his presidency. He further learned that Kennedy too had maintained a system. "Not as extensive apparently as President Johnson's, but that there were several hundred tapes of conversations that had occurred while he was president that had been taped and put in his library." Nixon had the Johnson system dismantled but installed a new one in February 1971 on Johnson's strong recommendation, "just in places where we conduct official business." This one would play a fateful role in his presidency.

Mr. Nixon's account was flavored with wordy descriptions of LBJ's persuasiveness and a syrupy recollection of a military medal ceremony he was happy to have taped. Nixon vaguely recalled having urged H. R. Haldeman to purge the tapes of all save historic content, but Haldeman recalled having received no such instructions. After Haldeman and John Ehrlichman were sent packing, Nixon, at the suggestion of General Alexander Haig, his new chief of staff, listened for the first time to all of the Watergate tapes involving conversations with John Dean. Keeping those tapes "was probably a good idea because the tapes in many respects contradicted charges that had been made by Mr. Dean . . ."

Finally he got to Butterfield and the question of destroying the tapes. "I considered it. Ah, but I felt, ah, that first, if I were to destroy them, it would be an indication that I felt there were conversations on there that demonstrated that I was guilty. I thought it would be an admission of guilt to destroy them."

Nixon related that after Special Prosecutor Archibald Cox had won the right in the Court of Appeals to subpoena nine crucial tapes, he had considered turning them over to Mr. Cox while destroying the rest in order to preserve "the confidentiality of all presidential communications." But again he claimed to have been influenced by not wishing to look as though he had something to hide. Moreover, ". . . I must admit in all candor that I didn't believe that they were going to come out."

In the end, Nixon cited those four main reasons for not destroying the tapes: he did not wish to act in the manner of a guilty man; he thought that if not exculpatory they would at least impeach the credibility of John Dean, his most dangerous adversary witness; he felt there were other means to protect the presidency against future encroachments by the legislative and judicial branches; and he never quite believed that the courts would compel him to part with the requested material.

His judgment proved faulty with respect to each reason. First, it was the tapes themselves that provided the most compelling evidence of presidential guilt. Second, rather than impeaching Dean—and despite his erroneously transposing a March 21 conversation with the president to March 13, 1973—Dean's memory proved superb, his note taking fastidious, and his "cancer on the presidency" advice to Nixon—cut it out!—the best the president received during his lengthy ordeal. Instead, Nixon persevered on his own disastrous course. As he spoke, I remembered Nixon's lame attempt to force Senator John Stennis upon the special prosecutor's office in order to monitor White House–supplied "summaries" of tapes subpoenaed by the office. The country wound up with the "Saturday Night Massacre," Mr. Cox's successor with the tapes, and the president—so worried about appearing as though he had something to hide—with another giant stride down the road to impeachment. That such a lengthy parade of lawyers, special counsel, speechwriters, senior staff assistants, attorneys general, and self-anointed defenders of the presidency were unable to get inside Nixon's head long enough to fix his course remains perhaps the greatest unsolved mystery of Watergate.

Though Nixon's wandering tour of White House taping systems was hardly the candid opening gambit I'd been hoping for, I stayed with the subject for one final question: "But looking back on it now, don't you wish you had destroyed them?"

Even as I asked the question, the entire bonanza of Nixon tapes had been impounded by an act of Congress, triggering a legal battle over controlling access that Nixon would eventually lose. The result would be unprecedented access by students of the presidency and other scholars to the inner workings of the White House, not simply with respect to Watergate and other alleged abuses of power but regarding relations with Vietnam, China, the Soviet Union, and others, and such important domestic issues as family assistance and civil rights. But from the vantage point of San Clemente, still weathering the long winter of his discontent, Nixon already knew the magnitude of his mistake, acknowledging that "if the tapes had been destroyed, I believe that it is likely that I would not have had to go through the agony of resignation . . . I think it would have been well, looking at it from our standpoint, to destroy them all."

It had taken a lot to get this small admission. In our bedroom at the far end of the hall, where Zelnick, Reston, Marv Minoff of Paradine, Peter Pagnamenta of the BBC, and David Gideon

Thomson of Polygram were watching on a 12-inch monitor, the morning's elated mood had begun to dissipate. Worries only mounted as Nixon began rambling again. He insisted that he had been judged unfairly because the tapes had been taken "out of context" and that publishing them in a way that showed his "very volatile" private side had undermined the essential dignity of the presidential office. He suggested that perhaps he should have destroyed the tapes, not for personal reasons but because "there's been a chilling effect, really, on the advice that future presidents are going to get" since the best advice is of a confidential variety.

"Ah," he said, "I know what cabinet officers and staff members and congressmen and senators and friends usually do when they come to see a president. Many times, of course, they want to be heard. They have some ax to grind, and that's very proper. Ah, at other times, they want to please the man. But above everything else, they want to be sure if they are heard, and they give advice, ah, and it proves to be wrong, that they don't take the blame. And that's why I always assured people . . . look here, you tell me what you think and I'll take the blame if it should fail, and I'll give you the credit and take a little of it myself if it succeeds . . ."

Clearly Nixon was nowhere near ready to confront the Watergate period with complete candor, but that should have surprised none of us. It was something he needed to build up to, Ken Khachigian had assured Bob Zelnick. "Don't be surprised if you encounter difficulties until the Watergate taping segment is under way."

On the set, I was disappointed but not overly distressed. I had gambled that an opening thrust into an area he knew a lot about but had not been a part of our day one agenda might produce a dramatic response. It had not. But little of what he had said would survive the final editing, so all that had really been lost was time. I had to believe he would be more responsive as we began discussing the day's agreed-upon subject, the last days of his presidency. I asked him when he had decided to resign.

The return to agreed-upon territory produced nothing beneficial with respect to usable material. Nixon produced a rambling, somewhat disjointed account of events that even in their barebones form would be hard to square with what we would come to learn had been going on inside the White House. Nixon said he decided to resign on July 23, 1974, following a telephone conversation with Alabama Governor George C. Wallace, which left him convinced that he would not capture enough support from southern Democrats to defeat an impeachment vote in the House. Weeks earlier he had been assured by his vice president, Gerald Ford, that the votes were there. But then three southern Democratic members of the House Judiciary Committee—Walter Flowers of Alabama, Thomas R. Mann of South Carolina, and Ray Thornton of Arkansas—all announced that they would support at least one article of impeachment. And on the twenty-third, Nixon received a call from Joseph Wagoner of Louisiana, the leader of a bloc of about 50 southern Democrats, who told him he could count on no more than 30 of those votes, which would leave Nixon with no more than 190 of the 217 votes needed to block impeachment.

Enter Chief of Staff Al Haig, who told Nixon he had just taken a call from a "Mr. Snyder," who had identified himself as an adviser to George Wallace. Snyder advised a call to the former segregationist pit bull, saying he could help turn Flowers and perhaps other southern Democrats. In desperation, Nixon called the man who had, along with Hubert Humphrey, been his opponent in the 1968 presidential campaign, only to find Wallace hard of hearing, remote, and unwilling to get involved. "I don't believe there's anything I can do to be helpful," he told Nixon. After a brief exchange of amenities, Nixon said, he hung up and turned to Haig. "I said, 'Well, Al, there goes the presidency.' "

As we shall see later, this account does not fully comport with what was going on at the time in the White House, where all appeared to be digging in for the climactic final battle. That changed abruptly, not on July 23 but rather on August 1, when presidential counsel J. Fred Buzhardt listened for the first time to the infamous "smoking gun" Nixon-Haldeman tape of June 23, 1973, in which the president ordered Haldeman to speak to the CIA and urge it to request the FBI curtail its investigation into Watergate. As I waded through Nixon's narrative, all of the fighting statements made between July 23 and August 8 stood fresh in my mind. If it was true that he had made his decision on the twenty-third, what purpose could those statements have served other than to place his desperate allies even further out on a limb?

But my problem at the moment was not authenticating the former president's account. Rather, it was to get him to respond to questions directly and with some concern for the constraints of time.

"Move in, tear the SOB to pieces," implored Zelnick in the monitor room.

"Move along. Move along, David," Birt countered in the production trailer.

I decided to press forward.

A question about his emotional state after he had decided to resign, however, elicited a veritable dissertation on his relations with President Dwight D. Eisenhower, their final get-together when Ike had been on his deathbed at Walter Reed Army Medical Center, and the tears Nixon had shed a day or so later when he learned that this crucial figure in his life was gone. Interesting stuff, had Nixon been able to describe it all in two minutes rather than ten, but even then not really pertinent to the subject at hand.

Nixon's final encounter as president with Henry Kissinger, unforgettably described by Bob Woodward and Carl Bernstein in *The Final Days,* was something else again. Kissinger had denied being the source of the story. Nixon certainly hadn't been. Yet no one but the two of them had been present. How accurate was the account? Had the two men wept and drunk cognac together from the same Courvoisier bottle they had used to toast China's invitation to the first summit? (In bringing En-lai's dispatch to Nixon, Kissinger had grandly announced, "Mr. President, this is the most important message a president has received since World War II.") Had Kissinger promised to quit if the Ford administration mistreated Nixon? (Kissinger would later tell me that from Nixon's embellished account of the Kissinger threat, "you would think the purpose of the meeting had been to discuss my resignation rather than his.") Did they fall to their knees in tearful silent prayer—the Jew who had escaped Nazi Germany with his family and the Quaker veteran of World War II? Yes to all of the above. Most likely Kissinger had designated a senior aide—Lawrence Eagleburger, we were told—to deliver Woodward and Bernstein an accurate account of the session with Nixon. And as Nixon had prepared himself both for our interviews and his memoirs, Diane Sawyer, formerly an aide to Ron Ziegler and soon to be a CBS and ABC news anchor, played a liaison role between the two camps, making sure their recollections of events both great and small were compatible. As Nixon told the story, however, it read like a scene ripped from the pages of a Danielle Steel novel.

"And Henry at that time, and I too," said Nixon, "became very emotional. He said, 'Well, Mr. President, I just want you to know,' he said, 'It's, it is a crime that you're leaving office. It's a disservice to the peace in the world which you helped to build, and history is going to record that you were a great president.' I said, 'Henry, that depends on who writes the history.' . . . And then he said, 'I just want you to know that if they harass you after you leave office, I am going to resign, and I'm going to tell the reason why.' And his voice broke and I said, 'Henry, you're not going to resign. Don't ever talk that way again.' "

Though this initial exchange shed little light, we returned to Nixon's relationship with Kissinger more than once in later sessions. As Nixon began to relax, his comments on Kissinger grew more colorful. Their relationship, laden with dislike, distrust, disdain, and disgust on the one hand and, on the other, admiration, appreciation, respect, and mutual need, is one of the more remarkable of recent times and one of the most worthy of further scholarly exploration.

In the period that remained, I asked a handful of questions relating to specific moments of Nixon's final White House days and departure. I recalled his admonition to the White House staff not to hate those who hate you because then "you destroy yourself." What a lovely moment that would have been for a bit of human introspection, I thought. I had to settle instead for a few homilies, beginning with the respect he felt for his White House barber, the legendary Milton Pitts, and culminating with the advice of his former football coach, Wallace J. "Chief" Newman, to get mad at yourself rather than your opponent when you lose.

He did leave some thoughts to come back to, some at later points in the interviews, others with the perspective of time. His greatest contribution as president, he said, had been achieving an honorable settlement to the war in Vietnam. Second was the opening to China, and third, the achievement of racial integration in the South—an achievement that few at the time would have associated with the architect of the "Southern Strategy" but that rings far truer today with the perspective of time. And while we were on the subject of history, I asked him, "What negative things do you think it will say about the Nixon administration?" He responded, "The primary negative will be that the Nixon administration engaged in political activities which led to the resignation. In other words, the bugging of the Democratic headquarters and so forth."

The moment of candor proved fleeting as Nixon quickly shifted to a complaint that he was being judged by a "double standard." After all, shortly after taking over the presidency upon Franklin D. Roosevelt's death, Harry Truman had pardoned fifteen members of his old Pendergast machine of Kansas City who had been convicted of "stealing votes." Of course, had Nixon pardoned the Watergate culprits as part of a full accounting process, he would have retired at the conclusion of his second full term.

As the discussion came to a close, my mind ran over the morning's proceedings. Nixon was indeed well prepared. He was also evasive, long-winded, and, at times, maudlin. Most of the day, I knew, would end up on the cutting room floor. I had a lot to think about in the next two days.

The return ride to Beverly Hills was not altogether without rancor. Still smarting from my having opened with the question on burning the tapes, Zelnick complained that I had yielded control of the conversation right out of the box and had never regained it.

"He had his way far too much," said Bob. "I know this is not what you were intending today, but you are going to have to challenge him much more."

"Yes, much more than I thought," I concurred.

Nixon had talked endlessly, blunting the thrust of any area under discussion, creating a risk that we would run out of time with much ground left uncovered and providing us with nasty editing problems down the road.

"Honestly, David," chided Bob, "if I hadn't been selling pencils on the street when you discovered me, I don't know what I'd do."

Birt was more sanguine. "Now, we learned some lessons today about how Nixon operates and about the need for David to be more assertive. Obviously Nixon can't recite *War and Peace* in response to every question. Now, I'm as concerned as you are about falling behind schedule, but I can promise you that if there are any cuts to be made, it will not be Watergate or other dirty tricks that gets the ax. Now let's talk about Vietnam."

Richard Nixon had, without a doubt, the most varied repertoire of parrying devices I had ever encountered. We had given him a soapbox—partly deliberately, partly productively. But he was the one who had built it into a platform. And somehow, we had to be able to prevent him from doing that.

March 25, 1977

Vietnam provided an opportunity for us to prevent Nixon from using the interviews as a one-dimensional podium from which to broadcast his administration's successes. Objectively, the policy had failed. In the little more than two years since he had brought the prisoners of war home and declared "peace with honor," the whole of Indochina had fallen to Communist governments. Vietnam had been reunited under Hanoi's control. Remnants of Washington's South Vietnamese ally had last been seen clinging to the skids of helicopters lifting off from the roof of its Saigon embassy. Laos, in case anyone noticed, was also Communist. Pol Pot, the Khmer Rouge butcher of millions, was in his glory in Cambodia. Before too long, he would be on the lam, tossed out by his Marxist North Vietnamese cousins, who were appalled by his inhumanity. So much for the dominos. Once one departed from Indochina itself, they simply ceased to exist. Thailand remained independent, as did Malaysia, Singapore, Indonesia, and the Philippines. Japan remained contentedly under the U.S. nuclear umbrella. Taiwan's precarious independence had become strikingly less precarious due to the advancement of "normalization" with the People's Republic of China.

This left us with one big question and several smaller ones. The big question was, given the untenable position Nixon had inherited from two Democratic presidents and the cost in both human and material terms of fighting on, why hadn't he simply pulled the plug on the venture shortly after taking office? The smaller question involved some of the actions that had been taken—the secret bombing of Cambodia and later the incursion into that country, the Christmas bombing, the puffy rhetoric—all undertaken to support a policy that might be termed "defeat without surrender."

I was comfortable with that and comfortable as well with the underlying moral premise—that it is wrong to continue a fight that is terribly costly to combatants and noncombatants alike when there is no reasonable prospect of achieving success. I must confess that I was particularly anxious to explore Nixon's actions in Cambodia, where the engagement was not inherited from prior presidents. A fine British journalist named William Shawcross had written extensively about the secret bombing, the incursion, and U.S. political support for a right-wing coup engineered by General Lon Nol, suggesting that the bombing had helped drive the Khmer Rouge deeper into Cambodia, expanding the Khmer Rouge's territory as well as its recruitment base. Clearly the Nixon administration must accept some responsibility for the resulting human tragedy.

"Mr. President," I began, "the whole area of foreign policy is such a vast one, but at the moment you took office, America's involvement in Vietnam was regarded by many as a disaster that was splitting American society at home in a very grievous way for what seemed to many an obscure or even mistaken reason. How did it look to you, though?"

The question did not demand brevity of reply, nor did Nixon provide any. He began with an anecdote that described his opening the president's bathroom safe the first night he was in the White House and finding only a single item: the briefing report for January 20, 1969. The United States had 538,000 troops in Vietnam then, and 14,000 per month were being drafted. In one of his favorite formulations—which I call "the path not taken"—Nixon said he could have gotten out, blamed his predecessors, and lost the war. But he didn't for two reasons: "One, because I think John Kennedy was right when, as late as July 1963, he said that if Vietnam were to fall, all of Southeast Asia would fall with it." Nixon also believed Lyndon Johnson had been right when he warned that "what was involved in Vietnam was not simply freedom for Vietnam, the chance to choose their own form of government, but the security of the United States, the credibility of the United States as a dependable ally, and, as far as our adversaries are concerned, as one that would take positive action to stop aggression." When he took office, Nixon conceded, he was critical of how the war had been conducted, but "I wasn't about to go down the easy political path, of bugging out, blaming it on my predecessors. It would have been enormously popular in America." But it would have been at "an enormous cost, eventually even to America but particularly to the whole free world."

I felt it important on this second day—particularly on as sensitive a topic as Vietnam—to demonstrate early on that Nixon's statements would not stand unchallenged. Thus I countered quickly:

> FROST: But wasn't staying there, I mean, that was also at a massive cost, wasn't it? In billions of dollars; in 138,000 South Vietnamese killed; half a million Cambodians; half a million North Vietnamese; and so on. That cost—it's a question of weighing one cost against another cost, isn't it? But you thought the cost was worth paying for what you got?
>
> NIXON: Looking at my term in office, yes. I think considering the kind of peace agreement we finally got in January of 1973, one which provided for a cease-fire, ah, one which provided for, of course, the exchange in return of our POWs. One which also provided for no violations in the future of South Vietnam's territory by the North Vietnamese, among many other things. I believe that having accomplished that, after those four long, tortuous years, was worthwhile. And that held for over two years. The cost, I agree with you, however, was very great. It was a close call, a very difficult call.

In October 1968, just before the election, Lyndon Johnson suspended the bombing of North Vietnam. In exchange, Hanoi was supposed to stop shelling cities in the South and respect the demilitarized zone, which demarcated the accepted temporary line between the North and the South. During the early months of his presidency, Nixon explored a variety of ways to achieve a breakthrough at the negotiating table. He suggested a mutual withdrawal of U.S. and North Vietnamese forces, then an in-place or "leopard spot" arrangement that would freeze the status quo for the North while still withdrawing U.S. forces. He tried without success to link improved ties with the Soviet Union to Moscow's help in prodding Hanoi to accept a reasonable compromise; later he would try the same thing with Beijing, also to no avail. "And so we got to the bottom line with them very early," Nixon recounted. "A line they hung to right until the last, until October the eighth, 1972. Whatever we did, mutual withdrawal, unilateral withdrawal, ah, nothing that we offered would they consider unless we agreed on our part to overthrow the government of South Vietnam and allow them to take over. And that we would not agree to."

One good reason for Hanoi's obduracy was the sense that South Vietnam was a ripening fruit that would fall into its lap before too long. The U.S. policy of "Vietnamization," announced by the president in the spring of 1969, had the first withdrawals of combat forces occurring the following autumn. Nixon purported to condition the withdrawals on the level of military activity, the behavior of North Vietnam, and the progress of South Vietnamese forces being trained by Americans. But in fact they proceeded on an orderly schedule, even through the big cross-border Spring Offensive launched by the North Vietnamese. In our exchange, Nixon defended the policy as part of a larger "four-legged" approach.

FROST: And when linkage wasn't working in North Vietnam, that was when Vietnamization became, more and more clearly, almost a substitute for linkage, was it?

NIXON: Our policy in Vietnam, as I indicated, was basically a three-legged, or even perhaps, if you call Vietnamization, ah, bring that into it, a four-legged position. Negotiate, military pressure, ah, going—working on their arms suppliers, the Chinese and the Soviet Union and any other countries, even the Romanians, who might have good relations with them, and then, finally, Vietnamization, providing the Vietnamese with the means to help themselves.

To support his policies during volatile political times, Nixon made some questionable judgments. Certainly one was Operation Menu, the secret bombing of a fifteen-mile-wide strip inside Cambodia, which served as a logistical and operational center and was suspected to house the headquarters from which military operations in South Vietnam were being directed. Cambodia's leader, Prince Norodom Sihanouk, clearly did not enjoy having his sovereignty disregarded by his powerful neighbor to the north and had privately suggested that the United States might go after some of the infiltrators "in hot pursuit." And since both the Cambodians and Vietnamese would know they were being bombed, one might have concluded that the bombing was more easily justified than the accompanying secrecy, particularly since it involved the deliberate falsification.

Nixon was unapologetic. "Mr. Frost, it was much better for the young Americans that weren't killed by the hordes of ammunition, rockets, and the rest, and the number of civilians that weren't killed. My responsibility was to protect these men." He argued again that the supplies captured and destroyed had saved American lives. "You know there was no Tet Offensive in 1969," he jibed.

"Well, there was no third world war either, but that wasn't necessarily the result of bombing," I replied.

Nixon did go public on April 30, 1970, with his decision to

launch a ground offensive on North Vietnamese forces inside the Parrot's Beak area of Cambodia, adjacent to South Vietnam. This was the famous "pitiful helpless giant" address whose bloated language sparked weeks of sometimes violent demonstrations and protest marches across the country. As the speech came just ten short days after a far more optimistic presentation, I asked Mr. Nixon to enlighten us as to what had changed. Instead of responding with specifics, he talked in general terms about a buildup of enemy supplies and then uttered some disparaging remarks regarding the quality of my research and promised to provide the kind of intelligence we were seeking. Khachigian would later hand Zelnick some pages from Nixon's soon-to-be-published memoirs listing a number of unfavorable developments inside Cambodia, including the fall of some provincial capitals, but nothing regarding an enhanced threat to South Vietnam. So much for the promised Nixon "intelligence."

I asked a long, emotional question about the United States' violation of Cambodia's "flawed neutrality," and how the administration's air and land incursions had changed the military landscape while its at least grudging support for Lon Nol had changed the political landscape. The ensuing slaughter of millions—did that weigh on Mr. Nixon's Quaker conscience?

"If I could, if I could accept your assumption, yes," Mr. Nixon allowed. "But I cannot accept your assumption. Ah, I don't accept it because I know the facts." The Cambodians had stayed out of the big war and kept their own civil war pretty much under control during their 1970–1975 period of "flawed neutrality." It was the North Vietnamese who had intervened and eventually overthrown the government. The Khmer Rouge "couldn't have lasted a month unless they had received enormous military support from the North Vietnamese."

After the wave of protests following the incursion into Cambodia, things quieted down. Nixon went to China. U.S. troop withdrawals continued. Draft calls fell. So did U.S. casualties. Congress considered legislation to convert to an all-volunteer service.

In the spring of 1972, Nixon considered how to respond to a massive armored charge by North Vietnamese mainstream units across the DMZ. Should he cancel a pending summit with Moscow, let the Russians cancel, or leave the South Vietnamese to their own devices? Militarily, the big options included bombing Hanoi, mining Haiphong harbor, and giving plenty of air support to South Vietnamese forces. His Treasury secretary, John Connally, offered the advice Nixon recalled for us: "Number one, the president cannot lose his war. Number two, that means he should go forward with this strong action. It should have been done long ago, but now he has got to go forward. Number three, I don't believe the Soviets will cancel, but in any event, under no circumstances should the president cancel. Put the monkey on their back by blocking the road to peace because of their involvement in Vietnam."

The summit came off as planned. On October 8, in Paris, the North Vietnamese agreed to the South Vietnamese government of Nguyen Van Thieu remaining in office, pending a more complete settlement. All U.S. forces would be withdrawn. The POWs would come home.

I felt it had been a good day's work. I had questioned Nixon firmly and responded to his challenges. John Birt pumped my hand in congratulations, and Zelnick seemed—for the first time in days—unashamed of the fact that we were working together.

The Nixon camp was equally enthusiastic. Brennan, Khachigian, and speechwriter Ray Price warmly congratulated Nixon and myself on the dynamics of the day's debate. Though their reaction caught us by surprise, we soon realized they too had reason to be pleased. Nixon had held his own. He had stated his convictions plainly and had not given way, even under aggressive interrogation. He had not stonewalled or lost his temper, nor did he seem too physically taxed by the day's energy expenditure. We were back on track.

I should add, however, that perspectives on the issues evolve with time. Shawcross's theory of how U.S. military action helped bring the Khmer Rouge to power has been questioned by many, including Shawcross himself. More likely it was the inevitable by-product of the Communist victory in Vietnam, a victory assured when Nixon first announced that the United States would withdraw and then tried to negotiate as though he were dealing from strength. His deal ending the war was where our next session would begin.

THE INTERVIEWS: DAY 3

March 28, 1977

Monday morning got off to a rocky start, when a camera malfunction set us back ten minutes. Nixon unhooked his microphone and excused himself.

"I'll wait in the room," he apologized. "You know when you get yourself ready for one of these things, you lose your edge if you let your concentration wander."

Once we got under way, however, the session proved to be one of our most memorable and significant. The period covered involved the October 8, 1972, negotiating breakthrough with Hanoi, the problem getting South Vietnamese President Thieu to accept the deal, the Christmas bombing, the final deal, and, within two years and three months, the final Hanoi offensive and the collapse of the South Vietnamese army. It was an adversarial session, not in the sense of our Watergate "trial" but because it involved two sharply conflicting theories of the case based upon a carefully constructed and intricate body of facts.

Earlier, Nixon had told us of an informal three-hour session during the June Moscow summit at which Soviet Communist

Party boss Leonid Brezhnev, Premier Alexei Kosygin, and President Nikolai Podgorny had taken turns hammering him to accept Hanoi's terms for ending the war. Nixon insisted he could be flexible, but not to the point of scuttling the Saigon government and forming the sort of coalition government Hanoi sought. Late in the visit Brezhnev called him aside and indicated that Podgorny might soon visit Hanoi to explain Nixon's determination, "and he also implied that they would try to contribute to a more forthcoming attitude in the talks on the part of the North Vietnamese." Nixon readily agreed to Brezhnev's accompanying request that he suspend air strikes in areas Podgorny would visit.

Between the summit and the October 8 "breakthrough" session with Hanoi, Nixon sweetened the deal. He pledged to have all U.S. forces out of Vietnam within sixty days of the release of POWs. Both sides would be entitled to replace assets in the South on a one-for-one basis. The United States, in addition, would pay $3 billion to Hanoi for purposes of reconstruction. On October 8, Hanoi responded by dropping its demand for the removal of the "puppet" Thieu regime, agreeing instead to a formula for internationally supervised elections. The war seemed all but over. "I'll never forget when Dr. Kissinger got back from Paris after that October eighth meeting," recalled Nixon. "Kissinger came in late in the day, and he said, 'Well, the president now has three out of three.' Because we'd always talked about, we wanted China, we wanted Russia, and we wanted Vietnam." Saigon had yet to weigh in, but there seemed little doubt of Thieu's endorsement of the deal that his Paris representative had accepted.

Could Nixon have had the same deal had he been so forthcoming at the outset of his presidency? Could the war have been shortened by three years if he'd made the same generous offer in 1972? "No way," replied Nixon. "Ah, no way, and I can base that on the secret negotiations and also the public negotiations that we had in

Paris and the hard line that the North Vietnamese took throughout that period."

Going back to Lyndon Johnson, said Nixon, Hanoi had always insisted not only on a U.S. withdrawal "but, as we withdrew, throwing Thieu and his government out of office." This was true, Nixon said, even when the administration used its "secret channel" in the spring of 1971 to offer a unilateral U.S. withdrawal within nine months—later reduced to six—in exchange for the return of its POWs. North Vietnam balked, insisting on the removal of the Thieu government as part of the deal. Why had Hanoi changed its tune? Nixon offered four reasons: the bombing and mining initiated by the United States in response to Hanoi's Spring Offensive, the developing U.S. relationships with Moscow and Beijing, the improvement in Saigon's fighting capability, plus a political factor: "It was clear then that I had an enormous lead in the polls and it was very likely that I would be elected in November. Kissinger played that hard." Settle now or face a man with no incentive to make concessions after November.

But with the deal seemingly clinched, Kissinger ran into trouble in Saigon. I asked Nixon whether Thieu's reaction amounted to "a minor eruption or a minor earthquake."

"Well, I would say that's typical British understatement when you say a minor eruption," said Nixon. "It was basically, a major eruption."

Thieu, as we were later to learn, hated the deal in its entirety, believing that permitting North Vietnamese forces to remain in the South was a time bomb bound to destroy his society. *"Un drôle de paix,"* as Thieu himself told me later and inscribed for me on a notepad. Nixon could not help out with that kind of big objection. "Some of them we couldn't possibly get. For example, his insistence that the agreement had to provide that all North Vietnamese forces would withdraw from South Vietnam, because the

North Vietnamese wouldn't even admit they had any forces in the South."

On smaller matters Nixon tried to help. Thieu wanted an immediate cease-fire supervised by international inspectors to prevent a North Vietnamese land grab in the immediate post-agreement period. And he wanted the DMZ to remain the effective border between the two Vietnams. Nixon offered backing on these two items. Plus he offered generous aid to Saigon and one-for-one replacement of needed military equipment. Finally, in a secret letter to President Thieu on November 14, Nixon promised to respond "with swift and severe" retaliatory action to any North Vietnamese violations. "What did you have in mind at that point?" I asked him.

"Well," said Nixon, "that was simply a letter that had at its purpose giving him the self-confidence he needed to sign the agreement."

I was somewhat incredulous: "But, you felt that you could get congressional approval for 'swift and severe retaliatory action' with full force?"

Nixon: "I—ah . . . with regard to 'swift and severe retaliatory action,' I felt that if the North Vietnamese, which they had so often done in the past, flagrantly and blatantly violated the agreement, that I could go to the country and to the Congress and get the support that was necessary to bring them into line."

I was skeptical, and it showed. After "peace with honor" and with the POWs safely home, could the president have gotten the go-ahead for new engagements, new deaths, new POWS? Nixon said he could have gone to the country and made his case as he had in the past.

I disagreed: "But the difference . . . the difficulty for you going on television at that point would have been . . . all the other times you went on television, although you were announcing military ventures, they were all part of getting out. At that point we would have been out and you have been going on television for support to go, maybe only with airplanes and bombing, but to go back in again. Do you think you could have?"

Yes, said Nixon. "Because the people, having supported the actions we had taken previously to get the peace, I think, would have supported what we had to do to keep the peace . . . I would have broken the case strongly. It would have been swift, it would have been massive, and it would have been effective."

"Un-bloody-believable," said Zelnick in the other room. "I can't believe that David got him to say that."

"The rest is a bonus," Birt whispered to no one in particular from his cramped space in the production trailer.

On set, I kept my eye on the ball and continued to press Nixon. Those were the carrots for Thieu, he explained. There was also a stick: "I indicated to him that if he didn't go along, we would have great problems in getting continued American support within the . . . and particularly within the Congress for aid to Vietnam." When that proved insufficient to bring along the recalcitrant ally, Nixon was forced to send "increasingly tougher and tougher" messages to Saigon.

Hanoi, meanwhile, was troubled by U.S. efforts to amend the still-secret accord. So it published the agreement and urged the United States to sign, triggering Kissinger's famous "Peace is at hand" declaration, which launched a sense of national euphoria in the United States. Nixon regarded that Kissinger move as a mistake because it put Hanoi, Moscow, and Beijing in the position of knowing the United States would have to deliver what it said was at hand. The statement, Nixon said, "boxed us a bit into a corner. . . . As far as the American people were concerned, after the election, after that statement was made, after we had reduced the bombing and so forth, ah, they wanted to put the war behind them."

When Washington refused to back down, the North Vietnamese came up with their own version of bait and switch. According to Nixon, "They began to hint and imply and then insist that the matter of the release of our prisoners of war, military prisoners of war, should be tied to and conditioned on the South Vietnamese government's release of civilian prisoners held in the South."

After meetings in November and December, Kissinger informed the president from Paris, "I can't get an agreement." He came home on the sixteenth. Two days later Nixon launched his massive "Christmas bombing" campaign against North Vietnam. During our discussion, Nixon objected to the term because the bombing had been suspended for eight hours around Christmas Day. However, on December 27, Nixon launched the heaviest bombing raids of the war. The following day Hanoi informed the United States that it was willing to return to the negotiating table. The more important objective—Thieu's consent—was also achieved. According to Nixon, "There isn't any question that it had, I believe, a salutary effect on Thieu, in that it indicated that the United States was going to still take action against the North Vietnamese" if they violated accords and understandings.

According to Nixon, Kissinger—despite contrary notions planted with liberal and media colleagues—supported the bombing. He even urged Nixon to go on television to rally support for it, advice Nixon rejected for fear of locking Hanoi into a rigid position. "Well, frankly, I was surprised and I was shocked, and he knew that," said Nixon, then typically concocting a rationale for Kissinger's self-serving behavior. It was all part of Kissinger's effort to leaven criticism of the administration in this liberal community: "Consequently, he often took a position with them that was basically more reasonable than the position that I appeared to be taking. He did not do it for the purpose of making points for himself, in my view, but because he wanted to keep his leverage with them. He wanted to keep his credibility with them."

Next to resignation, Nixon said, the December bombing had

been his toughest decision as president. Kissinger, he agreed, had picked a bad moment to whisper a dissent he had never expressed in White House deliberations. "In this instance, what concerned me, of course, was that we had very little support even from friends in the press," said Nixon. He said he had spent a lonely and depressing holiday in Key Biscayne, scorned as a practitioner of "government by tantrum," in the words of James "Scotty" Reston. Finally, Hanoi came back to the table. "And we didn't make a big point about the fact that, well . . . the bombing is what did it, heh, heh . . . the facts spoke for themselves."

I noted that had President Thieu accepted the October agreement, Hanoi would have been cooperative and the Christmas bombing would not have been needed. Nixon's rejoinder: "The new agreement was better than the October accord." Not his most persuasive moment, as he and Kissinger had been the ones to negotiate the October agreement. As many of those covering the story had suspected at the time, what made Hanoi announce the deal in October and come back to the table in January was the fact that the agreement was merely another step toward its ultimate victory.

I pressed Nixon further. Were the modest improvements to the October accords worth all the aircraft and airmen lost? "I mean, given a choice, wouldn't you say it would have been better if the October agreement, imperfect as it was, could've been signed, if you had your choice? Or was it better to have the Christmas bombing and a slightly different agreement?"

Nixon rambled and stumbled but conceded that even with the "substantial" improvements achieved, "I would say in retrospect, certainly from our standpoint, my personal standpoint . . . that we would have preferred to have taken the October eighth agreement." Not all the facts and figures were known at the time, but history has made the bombing look even worse. For one thing, we now know that there were no new North Vietnamese concessions.

The deal agreed to in October was the deal signed in January. As one Kissinger aide, John Negroponte, snapped, "We bombed North Vietnam to make them accept our concessions." The U.S. losses were also severe: twenty-five B-52s and twelve smaller fighters lost, forty-three U.S. dead, forty-four new POWs, all to improve General Thieu's mood. In the end, Nixon still had to tell him that, unless he signed the accord, he—Nixon—would sign it himself.

Watergate would soon cripple the Nixon presidency, erasing the sense of triumph of January 1973 and the president's declaration that "peace with honor" had been achieved. Congress was able to pass a bill denying funds for bombing operations anywhere in Southeast Asia as of August 15. And in October, Congress overrode a Nixon veto to pass the War Powers Act, giving the president only sixty days to conduct military operations not approved by the Congress.

When I asked Mr. Nixon whether he could have saved South Vietnam from the North Vietnamese assault of May 1975, he replied in the negative, claiming, "the blame has to be placed where it belongs, in the Congress of the United States." Not only had they precluded bombing and constricted the president's power to employ force, they had also denied President Ford's request for funds to replace the weapons and equipment South Vietnam had lost on the one-for-one basis called for in the now-fractured peace accord. "So—why did this happen? It happened because the Congress refused to grant President Ford's request to provide the funds for the South Vietnamese to defend themselves. If the South Vietnamese had had the necessary equipment, I believe they could have held on."

Was Nixon saying "that American assistance, military assistance, would have to be permanent if South Vietnam was to survive?"

Nixon responded, "Well, permanent as long as the Soviet Union was providing permanent assistance to the North, yes." Even under détente, the United States had the obligation to help its friends help themselves.

Later, Nixon would respond to my question as to whether it was realistic to expect the Congress to appropriate $3 billion to heal Hanoi's war wounds, with a long soliloquy on the magnanimity of the American people, their generosity after World War II, and the willingness even of returning POWs to lobby for the assistance program. The Nixon peace accord, then, rested on two political assumptions. First, that he could win majority support for an aid program of some $3 billion or more for North Vietnam. And second, that if the admittedly "fragile" agreement broke down, he could convince the Congress and the nation to endorse his military reintervention, at least as regards the use of airpower. Idealist or the ultimate pragmatist, I concluded, it would have been a remarkable experience watching his act had he served his full second term.

"Let's come back to the domino theory," I suggested. The intellectual legitimacy for Vietnam rested on the domino theory, articulated by Presidents Kennedy and Johnson and embraced by Richard Nixon. I reminded Nixon of the dominos, "the Philippines and Malaysia and Singapore and Indonesia and Japan." All or most of them were supposed to fall if Indochina fell. "Well, they haven't fallen, there's been no discernable effect. Were we therefore pursuing a whole chimera in being in Vietnam at all? Because the domino theory, Vietnam, Cambodia, and Laos—Indochina—fell in April, May in '75, and there's been no domino theory."

Nixon had no easy response. When he had visited the "dominos," mostly during the mid-1960s, they had been concerned about an American defeat. Maybe improved relations involving the United States, China, and the Soviet Union had taken some of the edge off the danger. Maybe staying as long as we did gave some of the putative dominos time to fortify their own governments. "I am not a domino player," said Nixon. "But as far as whether the domino theory has validity or not, I think it's too early to judge."

I reminded Nixon of an early White House speech in which he declared that failure in Vietnam "would spark violence wherever our commitments helped maintain peace—in the Middle East, as you've mentioned, in Berlin, eventually even in the Western Hemisphere. I think that was an overstatement, wasn't it?"

"We shall see," said Nixon, weakly suggesting instability in "countries like Argentina and the rest." There was also the problem of the Caribbean and countries falling under the influence of Castro.

The ultimate question for Nixon involved the human and material balance sheet reflected in his administration. "The cost, we've said, was probably, during your administration, 75 billion [dollars], 15,000 American lives, 138,000 South Vietnamese, half a million Cambodian, 590,000 North Vietnamese, and so on. As we look at that cost, and as we take into account the body of opinion that feels that was not so much a peace settlement as a piece of paper, of which, tragically, nothing remains today, do you still feel the cost was worth it?"

"It was worth it in terms of the period in which I had the responsibility," said Nixon. He again reviewed the Kennedy and Johnson policies and said he had rejected the easy option, "to dump on them." He continued, "All wars are terrible, but if there was ever one in which a country had motives that were decent, motives that were unselfish, this was that kind of war."

That the dominos did not fall very quickly, he conceded, "I suppose makes the case as to why we shouldn't have done it in the first instance. My point is perhaps we will be sitting here maybe two years from now. Who knows? Here or someplace else. Ask the question then. My point is that as far as the United States was concerned, the war in Vietnam was a terribly difficult war. . . . I hated every minute of it and would have done anything I possibly could to bring it to an end. . . . But my point is that as far as the war was concerned, at the time that I made the decisions, it was the only, in my view, right decision to be made."

We touched again on dominos, on good wars and bad wars, and on whether Vietnam, which ended badly, would constrain the United States and energize its foes in the years ahead. For a few years it looked as though it might, as the Soviets and their friends created mischief in places like Angola and Afghanistan. Later, in 1980, the United States elected a president who argued that the United States' weakness was providing the Russians with a "window of opportunity" that must be shut. A big buildup and some toughness of spirit could still condemn this "Evil Empire" to the "ash heap of history." Soon the United States tested the waters in Granada and Panama—places influenced more by the Monroe Doctrine than the domino theory. Stunningly, it won a bloodless victory in the Cold War. Sympathizers and critics saw an age of Imperial America in the works, while neoconservatives began plotting the "New American Century." Later, one U.S.-led coalition kicked Saddam Hussein out of Kuwait while another helped end Serbian domination over the former Yugoslavia's Muslims and Catholics. Then came 9/11, Iraq, and the war on terrorism, and suddenly people started recalling Vietnam. *Everybody's All-American* had suddenly become the *Headless Hegemon* on the world stage.

Back on the set, we were drawing to a close. I asked Nixon to respond to the theory of someone who had worked closely with him that the thought of losing had been so repellent to Nixon that he could not recognize Vietnam as a losing battle.

Nixon demurred: "I wasn't concerned about my quitting, I was concerned about America's quitting. I wasn't concerned about my credibility or prestige, or what have you, because the presidency is transitory. I certainly was concerned about America's credibility in the world. And I didn't want America to be a quitter." But Nixon's distaste for disloyalty came out a moment later. "Several members of my staff were bug-outters," he said. "But I didn't have them in the foreign policy areas, I had them working on things where they could have those views and not interfere with the conduct of foreign policy, and the very fact that one would make this kind of statement is an indication that I had him in the right position, by not having him in foreign policy."

I then asked Nixon whether the length of the war, the bitterness and polarization it had engendered, and the abuses of power it had spawned might lead to the conclusion that Nixon himself had been the last casualty of Vietnam. Some interviewees, confronted with a question not specifically expected but fundamental to what had previously been discussed, will reach deep into themselves and deliver a response from the soul. Not so with Nixon, whose guard is never completely down and whose motto might well be: If it's so goddamn important, it should be in my briefing book. Yet here Nixon's face became a mask of pain. He paused. He drew a breath. His lips tightened. And then he spoke.

"A case could be made for that, yes," he began. "Ah, there isn't any question but that in the conduct of the war, I made, ah, enemies who were, from an ideological standpoint, ah, virtually, ah, well, paranoiac, I guess. Oh, the major newspaper publisher told Henry Kissinger one night, right after the peace settlement, 'I hate the SOB's gut.' And, ah, naturally, ah, coming right after the time that we had been able to have the peace settlement, is an indication of how deep those passions ran. Ah, because that kind of attitude developed over a period of years. I mean, my political career goes be . . . over many, many years. But, the actions, and many of them I took with great reluctance but recognizing I had to do what was right, the actions that I took in Vietnam: one, to try and win an honorable peace abroad, and two, to keep the peace at home, because keeping the peace at home and keeping support for the war was essential in order to get the enemy to negotiate. And that was, of course, not easy to do in view of the dissent and so forth that we had. And so it could be said that I was, ah, if I, that I was one of the casualties, or maybe the last casualty in Vietnam. If so, I'm glad I'm the last one."

"Congratulations on a great session," said Zelnick en route to L.A. "You had complete command of the material. He got away with nothing in an area regarded as his long suit. My God, we fought the war for an extra four years just so Nguyen Van Thieu could rule by himself without a coalition government. Then he nearly wrecks the Paris Agreement that collapses with the first big Hanoi assault. And then to blame it on the Congress. I didn't think he'd have he balls to do that."

"Yes, indeed, I agree with all of that," said Birt. "Now we've got to get the story out. I don't know if it's Joe Kraft or Clay Felker, but somehow the word's gotten around Washington and New York that we're a bunch of softies and that Nixon's been having his way with us. Let's take John Stacks [of *Time*] and show him some of the clips from this morning's session. And Bob, on your next jog with Hal Bruno [of *Newsweek*], tell him how tough David was."

We had already survived one media challenge, the visit of Mike Wallace and his *60 Minutes* team. Some had predicted a "David in the lion's den" hatchet job. But Mike's producer, Harry Moses, spent enough time with us to conclude that we were in the midst of a serious piece of work. So Mike was the picture of amiability and the interview proved to be quite a bit of fun. One example, Mike asked me what approach I expected Nixon to take:

> FROST: I hope the approach he takes will be one of a cascade of candor.
>
> WALLACE: A cascade of candor from Richard Nixon? Is this what you expect?

FROST: No, it was just a phrase which I thought would appeal to you.

We did expect a cascade of candor on China, perhaps Nixon's greatest legacy, and for that reason adopted a different, distinctly nonadversarial mode.

THE INTERVIEWS: DAY 4

March 30, 1977

Our sessions on relations with China and the Soviet Union were more debriefings than interrogations, and for very good reasons. This was Nixon as he probably envisioned himself, doing big, dramatic things on a world stage for his country and the cause of world peace. And doing them well, perhaps as well as any president in history.

The moment was propitious for an American move. Sino-Soviet relations were bad and getting worse, with Beijing fearful that the Russians sought to dominate all Communist societies and, as a Eurasian power, wanted regional hegemony as well. Normal relations with Washington could be helpful—assuming the Taiwan issue could be finessed—on the theory that "the enemy of my enemy is my friend." Nixon's view was the mirror image of Beijing's.

With respect to the Soviets, there were several reasons to seek improved ties. The first was to get control of the arms race, preserving mutual deterrence against resorting to nuclear weapons, a concept threatened by the development of antiballistic missile systems as well as offensive systems with multiple, independently targeted warheads, each of which, in theory, could destroy ten or more enemy weapons. Second was the lack of communication between the superpowers, which could lead to a terrible misreading of intentions, as happened with near-calamitous consequences during the Cuban missile crisis. Third was the possibility of a regional brawl between great-power clients getting out of control, something that seemed always about to happen in the Middle East.

Nixon had one other item on his mind: finding a graceful exit from Vietnam. He was prepared to make major concessions, including a unilateral withdrawal while the North Vietnamese kept their forces in place. Hanoi was also insisting on removal of the Thieu government in Saigon. But it was dependent on China and, even more so, Moscow for its weapons and was thus far from immune to pressure from its big friends. That pressure was slow to come, the Russians promising late in the day to intervene.

Nixon's account of the secret steps leading to his historic visit to Beijing recalls no significant missteps. At various points, French President Charles de Gaulle, Pakistan's Yahya Khan, and Romania's Nicolae Ceauşescu served as intermediaries between Washington and Beijing. Walter Stoessel, the U.S. ambassador to Poland, fired up his old contacts with his Chinese counterpart. The latter brought a positive response from Beijing. Soon Henry Kissinger was secretly on his way to work with Chinese officials to set an agenda and draft a final communiqué. At first the Chinese insisted that the sole purpose of any visit would be to discuss Taiwan, but they would settle for an agreement to disagree written into the final communiqué, together with a U.S. restatement that it regarded Taiwan as part of China.

Nixon prepared for his trip by consulting with others who had met Mao Zedong and his foreign minister, Chou En-lai. The French writer and intellectual André Malraux warned that Mao would ask, "Would the world's most prosperous nation provide economic assistance to the world's most populous nation?" That never happened. Malraux also said, "When you see Mao, the thing that is going to impress him most about you is . . . your youth."

Once face-to-face, Mao did not ask for aid. Instead, Nixon felt

Mao was trying to take his measure, while leaving most of the substantive issues to En-lai. "As far as Taiwan is concerned," recalled Nixon, "the only reference was with their incomparable humor. He said, 'I see where your friend Chiang Kai-shek, ah, called me a bandit.' And after the translation I said, 'What does the chairman call Chiang Kai-shek?' Ah, and he said, 'Well, ah, I call him a bandit too.' And then Chou En-lai interspersed a comment . . . ah, he says, 'We just abuse each other.' Then they both threw back their heads and laughed."

Nixon's affection for the Chinese was both immediate and permanent. He admired their sophistication and subtlety as diplomats, their sense of justice, and their respect for struggle. The sentiment was mutual. At one point, Nixon recalled, En-lai spoke "very glowingly not of what I had done as president, or as vice president for that matter, but of the comeback . . . Mao also mentioned that . . . he'd read *Six Crises* and paid it a rather nice compliment, saying, 'You know, it wasn't a bad book.' And that from Mao is high praise. But in any event, what interested him most about it was not the achievements, not the crises, but the comeback, the comeback." Nixon had always viewed himself as the underdog. He was gratified to receive not only recognition but appreciation from the Chinese leadership.

The Chinese took the extraordinary step of inviting Nixon back in 1976, two years after his resignation in disgrace. By then En-lai was dead and Mao's days were severely numbered. Nixon said, however, that he was still "in charge of himself and he was still in charge in China." He also spent time with the new chairman, Hua Kuo-feng. To Kuo-feng, the Soviet hegemony issue was still critical. He told Nixon about a 1965 meeting between Mao and Soviet Premier Kosygin. "Our differences are going to continue for ten thousand years," said Mao to his host. Kosygin replied, "Well, Mr. Chairman, after these long discussions we've had and the reassurances we have given you, don't you think you could reduce that somewhat?" To which Mao responded, "Well, in view of the very persuasive arguments that the premier has made, I'll knock off a thousand years. Our disputes will continue for nine thousand years."

Despite the goodwill of these early meetings, however, the road to normalization with Beijing was not without its bumps. Nixon's old China Lobby friends treated the move as an act of betrayal. Conservatives like Ronald Reagan suspected it would undermine the morale of U.S. soldiers risking their lives in Vietnam to stop communism. The U.N. General Assembly promptly took the hint and bounced Taiwan, seating, to Nixon's fury, the delegation of the PRC. Four former ambassadors to Moscow, led by the respected Charles "Chip" Bohlen, warned the president that by cozying up to the PRC he would be sacrificing the opportunity to improve ties with the Soviet Union. During our discussion, however, Nixon maintained that the two efforts had complemented each other. The Russians "weren't about to want to have us have a new relationship with the PRC and a cooler relationship with them," said Nixon. "The Chinese game made the Russian game work, and the Russian game made the Chinese game work."

But it was the Chinese game that weighed most heavily on Nixon's mind. Comparing the two, Nixon said, "to have eight hundred million, and eventually a billion people, twenty years from now, with an enormous nuclear clout, and have no communication with them would be a danger to the United States, to all of the friends of the United States and Asia, and to the whole world, that we could not have. That's why the China initiative on its own was worthwhile. With the Soviet Union, it was more immediate. They were enormously powerful. They were equal to us. It didn't make any difference whether we were so many . . . we had too many . . . more warheads than they had or they had more missiles than we had. We both had enough to destroy each other."

Nixon spoke disparagingly of past summit meetings with the

Russians that had raised the "spirit" of the country only to produce disenchantment when no deals were reached. Ike had had his "Spirit of Camp David" with Nikita Khrushchev and Johnson his "Spirit of Glassboro" with Kosygin. "I wanted a summit in which we knew in advance that there was subjects and substance to be discussed, in which we could make real progress rather than to get together to shake hands, to tip glasses, to sign meaningless communiqués," said Nixon. Before they met in person, the two leaders made progress on "a step-by-step basis" until "several significant agreements" were reached. This time there was a deal—two major ones, in fact. One limited ABM systems to two sites, not anywhere near enough to prevent the nation struck first from inflicting "unacceptable damage" on the aggressor state. The second accord, the Strategic Arms Limitation Treaty (SALT), limited the numbers of land- and submarine-based offensive missile launchers.

When it came to nuclear arms control, Nixon and Kissinger had proven themselves conventional liberal thinkers. While conservatives had already begun to drift toward the sort of defensive system reflected in Ronald Reagan's Strategic Defense Initiative program, Nixon maintained that defense is nonsense: "If a nation is able to move forward and commit enough resources to the development of a thick nuclear system, ABM system, this would mean that it might gain enough sense of security that it might launch a first strike offensively. And therefore, that would change the balance of power in the world and increase the danger of war. In other words, you would no longer have the balance of terror. So by limiting defensive weapons, this new breakthrough, limiting them and practically eliminating them now, it means that option is not open to the leader of either of the two major powers, or any power for that matter in the world, which becomes a major power in the future. Because if you cannot defend yourself against a counterstrike, you are not going to launch a first strike."

Opponents of the deal—Senator Henry Jackson of Washington

chief among them—complained that the Russian limits on offensive launchers were slightly higher than the United States', a fact attributable to the higher Soviet production rate. Nixon argued that the numbers meant nothing: "Now, as far as numbers are concerned, there comes a point when they don't make that much difference. Because unless you can strike . . . knock out your potential opponent's ability to have a counterstrike, that gives you unacceptable damage. That, in other words, it kills forty, fifty, sixty, seventy million of your own people. And even were a leader insane, he cannot risk losing up to one hundred million of his own people in a second strike—that's not an option."

The agreements became the groundwork for Nixon's détente. Nixon: "Détente basically, to me, means talking rather than fighting. It means communication. It means agreeing in those areas where your interests are the same, disagreeing in areas where the interests are different, but having an understanding that you're going to talk about disagreements and you're not going to fight about them." Limiting nuclear arms appeared to lie within the framework of shared interests.

But the emerging détente was put to a serious test in October 1973, when Egypt and Syria—both Soviet clients that maintained no diplomatic ties to the United States—launched an attack against Israel designed to, at least, recover land taken by Israel in 1967. Months earlier Brezhnev, visiting San Clemente during Summit II, had spent three hours urging Nixon to persuade Israel to withdraw from the West Bank, the Sinai, and the Golan Heights. Nixon had demurred. He would not "impose a settlement." Withdrawal must be worked out in negotiations between the parties. Brezhnev was relentless, warning, "Unless the Israelis withdraw, the Egyptians and Syrians are going to attack and they are going to attack soon." Nixon responded that the United States would stand by its ally.

The attack came on October 6. Both the CIA and Israel's

Mossad failed to predict the attack, adding to Nixon's long list of grievances with his intelligence agency. ("I wasn't surprised to see our intelligence drop the ball," he said. "And I say this with all due deference to some of the good things they've done.") At first Israeli prime minister Golda Meir expressed confidence. Then the battle started going poorly and she phoned the president, asking for tanks, guns, and other supplies. When the Russians began an airlift of their own, her pleas became more desperate. The Pentagon, under Secretary of Defense James Schlesinger, dragged its heels for four days. Schlesinger wanted the cargo aircraft disguised as El Al planes to avoid political complications. According to Nixon: "I finally cut through all of the red tape and I said, 'Look, I mean, it isn't going to fool anybody.' I said, 'Second, what kind of airlift is it going to be?' " Pentagon officials said the plan was to send two or three C-130s. Nixon checked to see how many were available. He asked how many were in stock and was told the number was between eighteen and twenty-five. Nixon's next order: "I said, I want every C-130 that can fly to go in there. If we're going to do this, do it big, do it right, and do it quick."

The Israelis had soon turned the tables. Their conquered territories now out of danger, they sought to methodically destroy the Egyptian Third Army. Cairo, at Moscow's urging, called on the two superpowers to intervene between the warring armies. In short order, Kissinger received an invitation to meet with Anatoly Dobrynin in Moscow. "Needless to say," recalled Nixon, "I was anxious to have a cease-fire and I said, 'Go right ahead.' But with very firm instructions that we considered that this was a kind of conflict which should not be allowed to escalate; that we both had to restrain our clients; and that, ah, any action on the part of the Russian that, ah, escalated, would bring counteractions on our part." Kissinger was able to get an agreement for a joint U.N. resolution calling for a cease-fire. Within days, claims were made on both sides that the cease-fire was being breached. The Egyptians asked for a joint American-Soviet force to come in and keep the peace. Washington refused. In a conversation with Kissinger, Nixon recalled saying, "This is sheer madness. It may keep the peace, but it runs into the possibility of a big-power conflict. We can't do it."

It was then that Brezhnev sent a note that had, in Nixon's words, "an ominous sound to it." Clearly the Russians were talking about a unilateral troop deployment. Nixon responded by warning Brezhnev against unilateral action, placing U.S. forces—including nuclear forces—on heightened alert and persuading the Israelis to halt their offensive, thereby bringing the "Yom Kippur War" to an end. Egypt's president, Anwar Sadat, was unhappy with his advisers, his weapons, and the fact that Israel still held his land. Five years later he would switch patrons, recognize Israel, reclaim the Sinai, and entitle Egypt to $2 billion per year in U.S. military assistance. Three years after that, he would be shot. Egypt and Israel would for years maintain a "cold peace." Today that peace is threatened more by the rise of Islamic fundamentalism in the region than by issues specific to the Israelis and Egyptians.

Détente enjoyed a short half-life. Soviet and Cuban mischief in Angola got it off to a bad start. Pushed by Reagan during their 1976 battle for the GOP nomination, Gerald Ford dropped the term from his campaign vocabulary. Nixon argued that we were packing more weight on the doctrine than it could carry: "Détente should not be thought of as peace in our time or peace for time to come. Détente is simply a process under which nations with great differences agree, in effect, to discuss the differences rather than to come into constant confrontation with the possibility of a flare-up into war."

The 1979 Russian invasion of Afghanistan was a second dose of hemlock for the doctrine. Today it is a largely discredited policy associated with perceived U.S. weakness and Soviet belligerence and adventurism. Challenged by Ronald Reagan, the Communist society crumbled from within. Arms control—founded on a regime of sanity as we in the West define it—loses its heart when applied to cultures of death and suicide and weapons of mass destruction that can fit easily into a suitcase.

Even in 1977, however, Nixon had trouble defining any long-term benefits of the approach. I asked, "Do you think there's anywhere in the world since '72 and '73 where the Russians have actually behaved better on an issue because of détente?" The best Nixon could muster was an increase in the number of exit visas Russia granted emigrating Jews. I pressed further:

> FROST: Is there anywhere else in the world, though, that you would pick on as where the Soviets have behaved differently because of détente in the last three years?
>
> NIXON: Not that could come to mind, but ah . . . any questions that you have I'll be glad to answer on.
>
> FROST: No. I was asking you that. I couldn't think of one.
>
> NIXON: Well, of course, I was not there, and that . . . in that period of, ah . . . after August of 1974 and what happened in other parts of the world, I cannot really judge.

Nixon's China opening, however, still looks good. It seems more prudent now than ever to be on good terms with the heirs of Mao and En-lai as we move toward the close of the first decade of what may be their century.

THE INTERVIEWS: DAY 5

April 1, 1977

Richard Nixon hated bleeding hearts, particularly when it came to foreign policy. He and Kissinger were both superrealists. They could experience euphoria over a new relationship with Mao and En-lai even while the blood of millions of people savagely cut down during the Cultural Revolution was still fresh. They could pursue strategic arms control with Leonid Brezhnev, a ham-fisted Richard

Daley–type ward heeler if ever there were one. There were other aspects to the reality game: "tilting toward the Paks"—Kissinger's artful formulation—during their near-genocidal thrust against the seceding Bengalis and seeking to organize a coup to remove Salvador Allende, Chile's democratically elected Marxist president.

"He'll enjoy this session about as much as an appearance before the Ervin Committee," Ken Khachigian warned Zelnick as they discussed some of the question areas.

"Give me a break," Zelnick replied. "You can't expect all softballs like obstruction of justice and stealing Ellsberg's psychiatric files."

For me this was an important session. I had worked hard preparing myself on issues and events that were far more in Nixon's sphere of familiarity than in mine. Also, I needed to develop material interesting enough to get viewers to tune in to all four of our planned programs.

Allende first ran for the presidency in 1964. In September 1970, he received a 36 percent plurality of the vote. By September 15, CIA Director Richard Helms was in Mr. Nixon's office receiving instructions to try to prevent the Chilean legislature from confirming Allende as president. When that failed, the United States set about promoting a coup. While U.S. involvement in the 1973 coup that killed Allende has never been proven, it was certainly consistent with all Nixon's efforts. General Augusto Pinochet brought his iron-fisted authoritarian regime to power, apparently persuading himself that his extended rule was attributable to popular support. In the early 1990s, he restored the popular franchise to his countrymen, who promptly voted him out of power. He faced possible "crimes against humanity" charges in one forum or another until his death.

As Nixon and I spoke, of course, the brutal Pinochet regime was very much in business, squelching democratic institutions as well as people. To Nixon the realist, this was secondary: "In terms of national security, in terms of our own self-interest, the right-wing dictatorship, if it is not exporting its revolution, if it is not interfering with its neighbors, if it is not taking action directed against the United States, is therefore of no security concern to us. It is of human rights concern. A left-wing dictatorship, on the other hand, we find that they do engage in trying to export their subversion to other countries. And that does involve our security interests."

A valid, traditional argument, I thought, but not in the context of Allende, a left-winger operating within the confines of a democratic political structure. "Can you think of any other example," I asked Nixon, "where the United States in . . . recent United States history attempted to interrupt the constitutional process of a democratic government?"

NIXON: Well, it depends on what you mean by recent. Ah, well . . . you mean the last four or five years? No, I can't think of any.

FROST: Or even since the war—

NIXON: Since World War II—

FROST: Hm . . .

NIXON: Ah . . . well, I would suggest that, ah . . . in 1946 and '7 we didn't attempt to interrupt, ah . . . but we very, very strongly supported, ah . . . the [Italian] De Gasperi government . . .

"In fact," I declared, "what they have now with Pinochet is a right-wing dictatorship. What they had with Allende was a left-wing or Marxist democracy. It was never a dictatorship."

NIXON: Let's understand.

FROST: Was it . . . was it, though?

NIXON: No, I don't agree with your assertion whatever. I . . . oh . . . I would—

FROST: It was not a dictatorship, was it?

NIXON: It was . . . you said it was not a dictatorship, and my point is, Allende was very subtle and a very clever man.

I noted for Nixon that shortly before the fatal coup, the CIA had been saying Allende had no plans to abolish democracy and would likely be defeated in the next election. Not for the first time, Nixon unlimbered on the agency that had refused to participate in his cover-up of Watergate: "Based on the CIA's record of accuracy in their reports, I would take all that with a grain of salt. They didn't even predict that he was going to win this time. They didn't predict what was going to happen in Cambodia. They didn't even predict that there was going to be a Yom Kippur War."

But for the coup, Mr. Nixon suggested, Cuba and Chile could have become a "red sandwich"—two pieces of sandwich bread putting the squeeze on the rest of South America. I replied that Brazil, Colombia, Bolivia, Argentina, and others amounted to a lot of beef for these two tiny pieces of bread to control. Nixon was adamant: "We live in a world where at the present time the greatest threat to free nations is not from Communist nations, potential aggressor nations, marching over borders. It is not from Communist nations with huge nuclear armaments launching a nuclear strike, but the threat to free nations is through Communist nations, potentially, ah, aggressive Communist nations like the Soviet Union, like Cuba, for example, like Chile if Allende had stayed in power, burrowing under a border rather than over a border . . ."

"But there's two things there, surely," I countered. "One is that Cuba, which everybody would say is Communist in the traditional sense of the word, Cuba has been totally unsuccessful in its export of revolution or anything else since 1958, and Allende, ah, just didn't turn out that way. He turned out to be Marxist. He worked within the system for three years. He never attempted to introduce political repression. That only came later." My subject would not be persuaded.

Nixon said that for all the allegations, the United States had played no role in the coup that had killed Allende, that internal factors had been decisive: "I would have to say that [John] Foster Dulles was right . . . that the great failure of communism is that they seem irresistible in their ability to conquer a country either under or over a border, but they are totally inadequate and always fail in winning the support of the people of the countries they take over. Allende lost eventually . . . Allende was overthrown eventually not because of anything that was done from the outside, but because his system didn't work in Chile."

I tried one more time.

> FROST: If you had to choose a word to describe the Pinochet regime, ah, what adjective would you use . . . *brutal*?
>
> NIXON: Well, when they are brutal, yes. Ah, when they are dictatorial, I would say *dictatorial*. Ah, I would also have to, on the other side, indicate that they are non-Communists and that they are not enemies of the United States and that they do not threaten any of our neighbors.

If there was one thing I was learning from Richard Nixon, it was how to talk in circles.

The moment had come to discuss the Nixon-Kissinger relationship, my most delicious task of the project. Two more gifted, insightful, jealous, patriotic, secretive, self-promoting, reflective, manipulative, power-grasping public servants had rarely graced the same administration at the same time.

"When you first selected Henry Kissinger for the NSC, did you expect him to become . . . as much of an international star as he did?" I began.

"No. I didn't expect it. And I don't think he expected it either," said Nixon.

Nixon portrayed Kissinger as an intellectual who cherished his eastern establishment ties and loved hobnobbing with celebrities almost as much as he enjoyed conducting diplomacy. And though he conceded that Kissinger was a genius in his sphere, Nixon was careful to define the scope of their relationship. In recalling Kissinger's antagonistic relationship with Secretary of State William C. Rogers, Nixon observed, "This was a very painful thing for me because Rogers had been my friend. He was a personal friend. Henry, of course, was not a personal friend. We were, we were associates but not personal friends. Not enemies but not personal friends."

There was no doubt, however, that the two depended on each other. They adopted a kind of good-cop, bad-cop approach to diplomacy that served them well both at home and abroad.

FROST: It's a fascinating thing to note who did what and with which and to whom, and there was continually in the press this image of you as the hard guy, the—

NIXON: Or as the British say, the ornery guy—

FROST: The ornery guy, yes . . . And at the same time Kissinger in his conversations with the press and so on seeming to take a softer line, and was that deliberate?

NIXON: Quite deliberate. . . . Kissinger was an improvisor, he was one who believed in making startling plays and unexpected plays, and consequently he wanted a great deal of, ah, flexibility . . . I gave him that flexibility because I knew he would use it responsibly. But always under the proper direction.

Nixon described Kissinger as a man who would sometimes voice anguish, buffeted by second thoughts over decisions already made. For example, the Kent State shootings caused him to doubt the wisdom of the Cambodian incursion and suggest that the operation be called off ahead of schedule. Nixon steeled his colleague's will: " 'Henry,' I said, 'We've done it.' I said, 'Remember Lot's wife. Never look back.' " Kissinger's resolve would crack on many occasions, and each time Nixon would say, "Henry, remember Lot's wife." And that would end the conversation.

Kissinger also suffered huge swings of mood. While "cool and cold and controlled" in dealing with foreign leaders, he could go from elation to despair over diplomatic developments. "That doesn't mean he was emotionally unstable," Nixon allowed, practically begging one and all to reach the opposite conclusion. Yet he had to stabilize Kissinger's moods as Kissinger tackled issues such as Vietnam, arms control, and his "shuttle diplomacy" around the Middle East. As Nixon described it: "I tried to restrain elation, because I always know that, as Churchill once said, the brightest moments are those that flash away the fastest. And so that when you're up today, you may be down tomorrow. . . . Henry would feel highly elated by a conversation he'd had with Dobrynin, and then we'd have a bad development or a negative development and he would be greatly depressed. And I'd say, 'Well, Henry, the situation hasn't changed. We shouldn't have been as elated as we were yesterday, and we shouldn't be so discouraged today. Just keep plowing along.' " From time to time, Kissinger would threaten to resign, and Nixon would have to practice tough love—stroking Kissinger's ego while reminding him to keep his eye on the ball. So effective was Nixon's presentation that I half expected Henry to come bursting into the room in short pants, suspenders, and long socks, complaining that some big kids from State and the Pentagon had roughed him up and taken his soccer playbook. "Get a new one," Nixon would probably tell him. "And remember Lot's wife."

Kissinger hoarded power. His demand for secrecy meant that Secretary of State William Rogers was rarely current and thus rarely involved in major initiatives. "He wanted to be informed," said Nixon, "Well, Henry would come to me and say, 'I will not inform Rogers, because he'll leak.' And I'd say, 'Henry'—I must have told him this a dozen times—'Henry, the State Department bureaucracy will leak. It always has. It always will.' I said, 'But Bill Rogers will never leak.' " Henry never took Nixon's word, and as a result Rogers was rarely up to speed on Russia, Vietnam, and the Middle East. Nixon had to virtually beg Kissinger to keep Rogers posted on the China opening. "We had an argument about that," recalled Nixon. An argument Nixon won but that failed to enact any long-term change in Kissinger's attitude.

As his second term dawned, Nixon wanted to make John Connally his secretary of state. "But in this case, while Henry did not have a veto power—nobody can have a veto power where the president is concerned, any president—but while he didn't have a veto power, it was indispensable that whoever was secretary of state be able to work with Henry and Henry be able to work with him. In other words, I had gone through the Rogers-Kissinger feud for four years, and I didn't want to buy another feud with another secretary of state for the rest of the four years."

So Nixon went looking for someone who was as gifted as Kissinger but who would not threaten "his position of being the president's major foreign policy adviser." The science of cloning then unpracticed, the field rapidly narrowed down to one candidate: Henry himself. Nixon, Mr. Tough Guy, had been steamrolled by his own vetoless adviser. In photographs from Kissinger's swearing-in ceremony, Nixon appears as though he had spent the morning overdosing on castor oil.

Yet for all their personal wrangling, Nixon insisted that he and Kissinger shared a common view of the world and America's place in it. "As far as disagreements are concerned, I should emphasize, what disagreements we had were always on tactics and never on strategy," he said. Indeed, Nixon was even charitable regarding the many times Kissinger had been caught saying nasty things about his former patron, particularly a recent incident involving an open mike at an Ottawa banquet: "I can see exactly what happened in Canada. He runs into a lady who has a very low opinion of me and as Henry feels that really he's defending me and that the best way to defend me is to concede that 'Well, he's sort of an odd person, an artificial person' and so forth . . . The only problem was that he didn't think to turn the microphone off. On the other hand, I didn't turn it off either in the Oval Office on occasions, so I never held him for that."

Weeks later, just before our foreign policy show was aired, I was speaking on the telephone with Henry Kissinger in Washington.

"I suppose that this week you take off on me," he said. "I expect that I'm portrayed as a sort of neurotic genius in need of strong leadership."

"Henry, are you sure you haven't been bugging the sessions?" I said.

"Oh, no," said Henry. "I just know my boy."

From Kissinger, the conversation moved easily to Pakistan and the confrontation with India that had begun with the uprising in East Pakistan, aka Bangladesh. It was Kissinger who had inadvertently disclosed that the president had been on his phone every half hour urging a "tilt toward the Paks."

Nixon: "Well, I wasn't calling him on the phone every half hour, but he knew and he totally supported my view here that we had to do something to keep India from gobbling up Pakistan in direct violation of what Mrs. Gandhi's pledge had been to me when she had been at the White House November the fourth, a month before."

The air was thick with Washington-Moscow missives.

Nixon: "We ask . . . tell the Russians how important it is for them to restrain the Indians: They tell us how important it is for the Pakistanis to give up East Pakistan and, ah, there were a number of issues, but in any event, we both got across to each other our points of view, and our point of view was very strongly stated that we thought that if the Russians allowed their client, India, using Soviet arms, to destroy Pakistan, both East and West, that this would imperil our future relationship."

It was then that from a "totally reliable" source, the United States "learned that Prime Minister Indira Gandhi in a meeting of her cabinet had directed that a military force be put in place to attack West Pakistan." Nixon dispatched a carrier battle group to the Indian Ocean and sent another firm note to Brezhnev. A tepid Russian response was followed by more helpful activity once the carrier battle group was in place. The crisis abated.

I had been to Bangladesh shortly after the war and had seen evidence that the Pakistanis had slaughtered residents in numbers suggesting genocide—one to two million at least had perished—in an effort to crush the revolt. Nixon insisted that the United States could not permit a Soviet client to destroy a state that had been working cooperatively with Washington on several matters, including the China initiative. "But don't, let's not leave any impression that the government of the United States, that I personally or Kissinger or Bill Rogers or any of us condoned what the West Pakistani army was doing in Bangladesh. But that is an area with a terribly sad history. It's like the Mideast . . . there's going to be more of it."

FROST: What I'm saying is, but for the China initiative and those considerations we might have condemned it.

NIXON: We would have condemned it. Well, the point was, as far as that was concerned, who suggests that we approved it?

FROST: Nobody. Nobody . . . but we never spoke out against it, did we?

NIXON: We . . . but privately we did . . . of course . . . we took . . . we tried to give our best possible advice to Yahya and so forth . . . if you condemn the . . . to condemn it publicly, eh . . . would have served, in my view, no particular purpose at that point.

THE INTERVIEWS: DAY 6

April 4, 1977

As we walked across the kitchen for the start of our sixth day of discussions, Mr. Nixon turned to me and quite casually asked, "Well, did you do any fornicating this weekend?" For a moment, I could not believe the evidence of my own ears. Richard Nixon didn't say that, did he? He couldn't have. I must have heard it wrong. But no. One look at the stunned faces of the people nearest convinced me that I had indeed heard right. Even after thirty years, acquaintances will sometimes ask me whether he really said that and why. The answers are: Yes and I have no idea.

As we prepared to start our sixth day together, I had to smile at the clumsiness of Nixon's question. He was indeed trying to be one of the boys, and he got the word wrong. After all, lovers call themselves fornicators about as often as freedom fighters call themselves terrorists. And I knew that he did not really want to know the answer. So I said, "No comment. I never discuss my private life," as if at a mock press conference, and by then we were on the set.

I think Nixon probably did like to fancy himself one of the boys. Even his most loyal friends, however, remember the man as physically clumsy and verbally stiff. "I remember one of his jobs was to decorate soldiers—to pin medals on them," Brent Scowcroft said in an interview years later. "He was so physically awkward that he couldn't pin a medal on anyone. The first time he did it, he tore right through the jacket. So I had a clip-on version made. And still he was always fumbling. I was always picking up medals."

Our subject that morning was domestic policy and politics. We were steeped in the standard wisdom of the period, namely, that programatically, his record was weak and reflected his lack of interest in most domestic issues, while his politics—epitomized by his "Southern Strategy, his anticrime rhetoric, and his Supreme Court appointments"—reflected a pitch for the votes of bigots, racists, and political lowbrows generically described as "the Wallace vote." As I shall argue in the chapter reassessing Nixon and his presidency, this view inadequately treats Nixon's outstanding record on racial issues, the validity of street crime and violent political demonstrations as voting issues with the public, and his recognition that the enfranchisement of African Americans in the South would soon produce a true two-party America divided along liberal and conservative lines, with liberal Republicans and conservative southern Democrats finding no room at the inn. The new swing voters were essentially either middle- or working-class city dwellers or suburbanites who were attracted to Democratic social welfare programs but repelled by the Democrats' soft approach to law and order. In their classic work *The Real Majority,* Richard M. Scammon and Ben J. Wattenberg described this voter as "a 47-year-old housewife from the outskirts of Dayton whose husband is a machinist." To these two political scientists, this "social issue" could not be ignored. "The law-and-order issue today is essentially a civil libertarian's issue, and the question that must be asked is: What about the civil liberties of hardworking, crime-scarred Americans, black and white, many of whom happen to be Democrats?"

But in 1977, my team and I had begun to see Nixon's domestic policy as an outgrowth of his foreign policy decisions. Influenced by the writings of *New Yorker* editor Jonathan Schell, we wondered if Nixon's need to rally a domestic constituency behind his Vietnam War policies had led him to promote a hollow domestic agenda that lacked muscle but appealed to targeted voter groups.

"I know it means darting in and out of a great many different subjects," said Zelnick during our brief lunch break. "But with any luck, Nixon won't realize quite what we're getting at." Indeed, I'm not sure he ever did.

On set, the session got off to a rocky start as Nixon and I battled inconclusively over crime statistics, mainly in the District of Columbia, employing statistics that were inapposite and, in his case, unintentionally erroneous. Nixon became defensive, hostile.

NIXON: I won't contest your figures, there may have been an increase, but I'm not going to sit here and take the suggestion from you that I deliberately misled the American—

FROST: No.

NIXON: —people with regard to what we were going to do. Because I didn't.

FROST: No.

NIXON: If you want to make that charge, you go right ahead and do it, but I'll deny it.

FROST: No. I wasn't saying that on that particular issue that you did.

NIXON: Why don't you say what you meant, then?

FROST: Well, let me repeat it. What I said was that "Maybe you raised false expectations, false hopes, that there could be a simple cure for the crime rate."

NIXON: I didn't.

He was better at dealing with policy and principle, as when he quoted a 1968 Walter Lippmann essay: " 'The balance in our society has shifted dangerously against the peace forces, against mayors, governors and courts and police,' " quoted Nixon. "And that was true then. It was true in 1968. We tried to right that balance."

As a candidate, Nixon had been critical of the Supreme Court and pledged to appoint "strict constructionists" to the bench, a philosophy more easily stated than defined. His appointment of the pedestrian Warren Burger as chief justice sailed through. Clement Haynsworth was voted down on spurious ethical grounds, G. Harrold Carswell because he was an intellectual mediocrity. Nixon admitted that Carswell had been a mistake "basically because of his lack of experience . . . His legal credentials, as I look at them in retrospect, were not equal to those of Burger, of Blackmun, or Rehnquist and Powell."

At the time, Nixon blamed the Carswell vote on the unwillingness of the Senate to swallow a southerner. That was not true, as the subsequent confirmation of Lewis F. Powell, Jr., a Harvard-educated Virginia patrician, would show. What Nixon didn't fully comprehend was the dependency on the Warren-era court to correct the social and political inadequacies of the other two branches and the states. The Court, during a span of fifteen years, had become the forum of default for righting the wrongs of race discrimination, archaic political districting, unequal legal treatment of the rich and the poor, excessive police conduct, and thinly veiled assaults on freedom of speech, the press, and religion. Not even Nixon could change that overnight. Of his four confirmed appointments, only William H. Rehnquist, later the chief justice, could qualify as a conservative activist. Harry A. Blackmun was an outright liberal, Powell a reasoned eclectic, and Burger pretty much a practitioner of self-restraint.

"You can't peg these men," offered Nixon generously. "And I never asked them to be pegged. Ah, in the case of Powell, in the case of Rehnquist, in the case of Blackmun, in the case of Burger, ah, when I called them in to tell them I was going to name them, I told each one of them, starting with the chief justice, I said, 'You'll never hear from me again except when we meet on social occasions, and particularly, you'll never hear discussed any matter that indirectly or directly involves the Court.' And that was always my view. I said, 'You call 'em like you see 'em.' " The justices, apparently, took Nixon's words to heart.

"When did you think of the Southern Strategy?" I asked Nixon.

"I didn't think of it," he replied. "The Southern Strategy basically was a tag that was placed on the administration, on my candidacy, because I happened to win the southern states. I thought of a national strategy."

He returned to the subject moments later: "I wanted to bring the South back into this Union. Back into this country. That's why I appointed southerners to the Court and to other positions as well. And that is why when it came to the difficult problem of getting rid of de jure segregation, that instead of making a grandstand play out of it, I did it quietly, with persuasion and with great effect."

Nixon did in fact achieve an end to nearly all de jure segregation pretty much as he claimed. He began by giving the South a little breathing space by working through the courts rather than issuing federal agency decrees. But once the Supreme Court declared that the time to dismantle dual systems had come, Nixon moved with alacrity, inviting black and white community leaders to Washington, sitting them down with his distinguished labor secretary, George Shultz, and letting them work out the modalities of change. In terms of the number of students south of the Mason-Dixon Line soon attending schools in unitary districts and with a minimum of violence, Nixon's accomplishment ranks as a minor miracle, or perhaps a major one.

Nixon: "By getting these top leaders in, they did the job, and it was a splendid job. Now, and this is one figure we can't quarrel with at all: sixty-eight percent. When I came into office . . . and this is, you have to understand, is fifteen years after *Brown v. Board of Education;* after eight years of the Kennedy administration; the

Johnson administration, who had played so much lip service to civil rights and desegregation and the rest. And yet, when we came into office after all of that rhetoric and all of that lip service, sixty-eight percent of all black children in the South were going to majority black schools. And now it's eight percent . . . are in majority black schools. I don't mean we get all the credit for that. In fact, we get, perhaps, just a modest amount . . . there'd been too many promises made; too much talk and too little action. And we had action." Indeed, he did.

Nixon believed in nondiscrimination rather than forced integration through compulsory busing, scatter-site housing, and other artificial means. He and Shultz worked out the so-called Philadelphia Plan, whereby previously all-white construction unions were forced to admit blacks according to their proportionate representation in the available labor force, a "quota system," arguably justified as a logical remedy for specific acts of past discrimination. Under Nixon, the Office of Federal Contract Compliance (OFCC) "encouraged" firms contracting with the government to establish specific minority hiring "targets" in order to reduce the vast disparities in representation.

The former president seemed particularly proud of the way he had dealt in black businesses to compete for the billions of dollars in federal programs: "We came to office . . . only eight million dollars a year of those federal contracts were going to black businesses. By 1972, two hundred fifty million dollars a year were going to them."

Interestingly, as in so many other areas, Nixon argued that his domestic goals were best pursued under the radar and behind closed doors. "It's easy to demagogue this issue," said Nixon. "And I'll be very candid . . . Republicans have demagogued and Democrats have demagogued it ever since the Civil War . . . trying to get the black vote." But tough rhetoric had only inflamed the South, and Nixon felt it was time for a quieter approach. His unwillingness to condemn southern foot-dragging publicly cost him dearly with the press, as well as with the "civil rights leadership." Nixon: "They were all, ah, generally opposed to what we did. Not to what we did but to what we did not say." But Nixon maintained that his tactics got the job done. "Going out and, ah . . . hitting the South basically over the head . . . would have been counterproductive."

Thirty years ago, preparing for the interview on domestic issues, a Zelnick memo to me read as follows: "There is little question that Richard Nixon knew less about domestic than foreign policy, that he cared less about it, and that he accomplished strikingly little in the domestic realm during his presidency. Welfare reform, consolidation of the bureaucracy, tax reform, national health insurance, and some form of assistance to the nation's declining cities remained programs to be written by Jimmy Carter on a slate left virtually clean by Richard Nixon. Energy policy remains unsettled. Nixon left a modest revenue-sharing program and a more conservative Supreme Court, but little else. Even his new economic game plan of August 1971 seems undramatic now, created as it was to combat figures of 5.5 percent unemployment and 4.5 percent inflation which already appeared as almost an economic paradise."

That assessment may be harsh. Nixon's policy on race relations was far more nuanced, comprehensive, and intelligent than many realized at the time. His formidable environmental record suffered more from an absence of self-promotion than content, and his insights into what it takes for a Republican to campaign successfully in the ethnic Democratic North helped color the political map for a generation to come, though during his period in office, he seemed unwilling or unable to transfer to his party's House and Senate candidates the magic he worked for himself. Then, as Watergate grew big, the magic faded.

THE INTERVIEWS: DAY 7

April 6, 1977

John Birt had some necessary business to attend to in London and, upon his return, sat down to view the session he had missed. I told him my sense had been that Nixon had never known what we had been driving at and his answers had at times lacked focus. After viewing the tape, he concurred that Nixon had seemed oblivious to our theme. "The trouble is that I was oblivious to it too, as will be the audience." John recommended making short shrift of our remaining civil rights discussion and then moving right along. "I suspect that very little of what I've seen of Monday afternoon will survive editing."

I did not fully appreciate it at the time, but as we headed toward the last of our pre-Easter taping sessions and the period of round-the-clock preparations for Watergate, the project was experiencing a decline in momentum, a sinking of morale, and a shaking of self-confidence that forty-eight hours later would envelop us like smog settling into a coastal canyon. Fatigue certainly played a role in this, affecting not only our editorial people but also directors, crews, editors, and administrative assistants. But there was also a far deeper concern among John, Bob, and Jim that most of the accumulated material was falling into one of two categories: not sufficiently interesting to make air or subject dominated by Richard Nixon. Marvin Minoff had a term for the former—"too PBS." Zelnick would sometimes mutter to himself, "bartender or cabbie," two career paths he felt confident would remain open to him should our enterprise be judged a failure.

But on Wednesday morning, during the drive to Monarch Bay, such undercurrents were still not readily visible. Our primary subject for the day would be Nixon's approach to domestic unrest, and both John and Bob had similar points they wished to make. "Nixon knew that by making the decisions he did to continue fighting the war until he could obtain peace on his terms, the trauma in America would continue," said Zelnick. "Then, later in the October–November period of 1969, he realizes that he can either change his policy and quiet dissent or take the battle to the dissenters by mobilizing the so-called silent majority. The direction he chose was cold and calculated."

"In effect," Birt interjected, "he chose a policy of divide and rule, and there was nothing casual or accidental about it. Indeed, it was not even limited to the war issue. You've already been exploring its ramifications in areas like crime and civil rights. It was as fundamental a characteristic of Nixon's leadership style as anything one can mention."

"But what is the alternative?" I argued. "Surrender to the will of a dedicated minority?"

"In a democracy, that may very well be the alternative," said Birt. "At its best, democracy is a constantly shifting process of accommodation to majority and minority sentiment. The convictions of a minority may often be permitted to prevail if they are held with greater passion than those of the majority. In a free society, intensity of feeling is often as important as the mere numbers that would be reflected through the taking of a plebiscite."

"I see your argument," I said. "But I don't feel it. I don't think it's a strong case or a particularly desirable inevitability."

"One other thing," said John. "You spent a lot of time in the United States during the Nixon presidency. If you don't recognize the picture he draws of the America he saw from the White House, don't hesitate to tell him so."

Nixon was prepared for my questions on his mobilizing public sentiment against the war dissenters: some of them were violent. Also, the other side might otherwise conclude that public opinion in the United States had changed and the majority now wanted withdrawal: "In other words, are we going to have a situation where this war would be lost in Washington as the French lost in 1954 in

Paris rather than in Dien Bien Phu?" Accordingly, "We had to convince the enemy that this very loud but minority group of dissenters was not all of America." Thus his "Great Silent Majority" address of November 1969.

Nixon dismissed any suggestion that the speech broke with campaign promises to "bring us together," speaking quietly and "listening to the voices of the heart and those that were left behind." Nixon: "I tried to be what I am . . . and that is, I do the job that has to be done and I do it as fairly as I can and if it requires being firm, I'm firm. Ah . . . if it requires persuasion, I persuade. If it requires, ah . . . in some cases the threat of using the law, the . . . or what I call the stick in the closet, in order to get people to comply, ah, I go that way. That's the way I am."

The speech was enormously successful both in terms of the immediate response and its impact on the polls. Gallup reported Nixon's approval rating as standing at 68 percent. At the same time, Nixon unleashed his own "Nixon": Vice President Spiro Agnew and his alliterative attacks on "middle-aged malcontents," the "nattering nabobs of negativism," and, far more significantly, the eastern establishment media with its anti-Nixon bias. "Was the negativism and all those alliterations, was that your idea?" I asked Nixon. Nixon: "No. . . . As a matter of fact, what editorial suggestions I made, ah, were frankly trying to make it not so cute." Not that Nixon was displeased to see the media given a dose of their own medicine. "Whoever is the leader of this country . . . has to defend himself and defend his policy and to use every resource that he can to get his point across or otherwise these people, who were not elected by anybody and who are paid very well . . . most of them particularly in television by their superiors . . . they run the country."

Though he had long served as the press's piñata, Nixon wasn't sure where the animosity came from. His onetime nemesis Dean Acheson had told him what the problem was: "You want the press to love you. And they won't because you're not a lovable man." More seriously, Nixon said, "The reason they were against him [Johnson], most of them, and the reason most of them were against me . . . was not because we were not lovable, although I don't think either of us is in the sense that they want. But . . . because of what we stood for. They were against what we stood for. And, ah . . . so that's the way it is."

Nixon denied that his rhetoric had been particularly harsh. One opponent had compared the United States in Vietnam to Hitler's extermination of the Jews. Another proclaimed him a "maddened tyrant." R. Sargent Shriver, George McGovern's running mate, called him history's greatest bomber, "even greater than Julius Caesar," who—Nixon noted—lived a thousand years before gunpowder was invented. He recalled a meeting with George Christian, Johnson's press secretary. Nixon suggested "that President Johnson is going to be real pleased when he finds out that now they're calling me the number one bomber. And George Christian said, 'Oh, don't be too sure, 'cause you know LBJ, he never likes to be number two.' "

We were going nowhere. Nixon's answers were expansive, but I couldn't see much of the material surviving the editing room. It was time to move to more controversial territory. Why had he approved the Huston plan, which included the "black-bag jobs," or burglaries?

"In the Huston plan it stated very clearly with reference to the entry being proposed, it said very clearly, 'Use of this technique is clearly illegal, it amounts to burglary,' um . . . however, 'It is also one of the most fruitful tools and it can produce the type of intelligence which cannot be obtained in any other fashion.' Um . . . why did you approve a plan that included an element like that . . . that was clearly illegal?" I asked.

In a rambling, almost incomprehensible response, Nixon talked about the dilemma a president faces in weighing national security concerns over personal liberties. In this instance "black-bag jobs" had been necessary in order to collect information on two violence-prone domestic groups, the Weathermen and the Black Panthers. He acknowledged that neither had received material foreign support but said the number of bombings, bomb threats, assaults on policemen, and airplane hijackings had created a dangerous domestic situation. "Under the circumstances," he said, "I felt that we had to coordinate these activities and get a more effective program for dealing with first, foreign-directed, ah . . . espionage, ah . . . or foreign-supported, ah . . . subversion. And in addition with domestic groups that used and advocated violence."

His order approving the plan remained in effect for less than two weeks, until FBI Director J. Edgar Hoover told Attorney General John Mitchell he could not go along. Nixon speculated that Hoover, in his final years, became fixated with his reputation in the broader community. Without the FBI, there was no agency to conduct the program—"the whole house of cards fell"—so Nixon rescinded his order and the Huston plan died.

"And a month later, you know what happened? Well, they blew up, when I say that they . . . one of these groups blew up a building at the University of Wisconsin, Madison, and, ah, a twenty-five-year-old student was killed and two others were injured . . . and the property damage, of course, was in the millions. If we'd had the plan into effect, maybe he'd be living."

"In retrospect," I asked, "wouldn't it have been better to combat that crime legally, rather than adding another crime to the list?"

On this first specific pass, Nixon dodged the question in not untypical fashion, suggesting that other presidents might also have violated the law, Lincoln by suspending habeas corpus and, later, shooting down antidraft protesters, Roosevelt by the Japanese internment program. "I considered that the president had not only the power but he had the responsibility in this instance to put the safety of citizens, ah . . . above, ah . . . the legal technicality that was involved. But on a very limited basis," said Nixon. All of Nixon's examples, of course, were inapposite. The suspension of habeas corpus had been a public action reviewable by the courts. The antidraft protesters had been on a murderous rampage. The Japanese action, to the country's eternal shame, was approved by the U.S. Supreme Court.

FROST: So in a sense what you're saying is that there are certain situations, and the Huston plan, or that part of it, was one of them, where the president can decide that it's in the best interest of the nation or something and do something illegal?

NIXON: Well, when the president does it, that means that it is not illegal.

FROST: By definition?

NIXON: Exactly, exactly.

A flurry of questions stormed my head. How far would Nixon go? Without betraying any emotion, I pressed my subject further.

"So that the black jobs that were authorized in the Huston plan—if they'd gone ahead—would have been made legal by your action?"

"Well, I think that we would . . . I think that we're splitting hairs here."

Burglaries were illegal, but if a burglary were undertaken because of an expressed policy decided upon by the president for national security reasons or issues of domestic tranquillity, "then that means that what would otherwise be technically illegal does not subject those who engage in such activity to criminal prosecution." What about murder? What if the president ordered an assassination? "No . . . absolutely not . . ." replied Nixon. And a moment later: "the Huston plan, as you know, is very carefully worded in terms of how limited it is to be."

FROST: But no, all I was saying was, Where do we draw the line? Why shouldn't the same presidential power apply to somebody who the president feels in the national interest should murder a dissenter? I'm not saying it's happened. I'm saying what's the dividing line between the burglar being liable to the criminal prosecution and the murderer?

NIXON: Because as you know from many years of studying and covering the world of politics and political science, ah . . . there are degrees, ah . . . there are nuances, ah . . . ah . . . which are difficult to explain, but which are there.

Within minutes, Nixon had waded even further into the muddle and was ruminating about how nice it would have been to have assassinated Hitler before he launched his own extermination campaign against European Jewry. But by now I almost didn't care how far he strayed or how self-serving his arguments became. I believed he had presented us with a stunning picture of his mind-set by advancing the proposition that the president has the inherent power to violate the law and, by so doing, to purge the entire transaction of its unlawful character. I sought to underline the unique circumstances with which we were dealing.

"In fact, is there any single case at all where any former president has personally approved black-bag jobs?" I asked.

"I can't speak for any of them," replied Nixon. "None have ever indicated to me that they have."

In typical Nixon style, he then launched into an utterly irrelevant distraction on covert action targeting foreign embassies.

After a break, I followed up on Birt's earlier suggestion and asked Nixon which parts of the country he had been referring to when he described part of the country as having been in a "virtual state of revolution."

"Primarily in the . . . in the major cities in the eastern part of the country," said Nixon. "And primarily in those areas against education . . . near to educational institutions, which might have a more liberal activist element in them."

"Thinking back to that period, it seems to me that's a massive exaggeration," I countered. "I don't remember that time in America, in the East, there being a virtual state of revolution . . . revolution implies so much, I thought."

Nixon didn't handle the challenge very well, lumping together the excesses of peace demonstrators with a bomb discovered in Detroit, a rise in airplane hijackings—nearly all by foreign nationals—and thousands of assaults against police officers, most by ordinary street criminals. He did much better responding to my suggestion that his war policies were responsible for most of society's violent divisions, recounting the urban riots that had begun in 1964 and run through 1968, the year of his election. The war had contributed, he acknowledged. But he rejected the notion that he had sought to divide society in order to isolate the war critics.

> NIXON: But why did I go to China, then? Why did I go to Russia?
>
> FROST: Well, I think—
>
> NIXON: Why did I take the positions I did with regard to chemical and biological warfare? Why did I take the position for family assistance?

As to taking on the dissenters at home, "I knew that in order to get the enemy to take us seriously abroad, I had to have enough support at home, that they could not feel that they could win in Washington what they could not win on the battlefield. And I had that support."

Nixon wanted to talk about the press and in particular two of its high-flying representatives, Bob Woodward and Carl Bernstein of *The Washington Post.* Their best seller *The Final Days* had presented both Nixon and his wife as heavy drinkers whose marriage had become burdened by unresponsiveness and a long-standing lack of intimacy.

According to Nixon, "The greatest concentration of power in the United States today is not the White House. It isn't in the Congress and it isn't in the Supreme Court. It's in the media. And it's too much." He spoke of large newspapers owning large newsmagazines and, of course, the three networks. He called the Supreme Court decision requiring actual malice in cases involving public figures "a license to lie."

"I don't want them repressed," he declared, "but believe me, when they take me on, Democrat or Republican, liberal or conservative, I think the public figure ought to come back and crack 'em right in the puss."

He saved his Sunday punch for the two *Post* reporters, but without mentioning their names. It came in response to a question on Mrs. Nixon's health. She had suffered a stroke several months earlier—purportedly after reading the Woodward/Bernstein book—but had battled back and was now assured of full recovery.

Nixon: "I've mentioned the stories that have been written, and some written by some book authors and so forth, which reflected even on her, on occasion, and what her alleged weaknesses were. They haven't helped, and as far as my attitude toward the press is concerned, I respect some, but for those who write history as fiction on third-hand knowledge, I have nothing but utter contempt. And I will never forgive them. Never."

We wrapped for the day, but I was intercepted by Jack Brennan before I reached the door. "This is the greatest material yet," he said, smiling broadly. "If you cut this out of the show, I'll put out a contract on your head."

"Jack," I said sincerely. "I wouldn't dream of it."

In the light of all that has happened since, it is hard to believe that our discussion of the Huston plan took place thirty years ago or that Mr. Nixon's approval of it—lasting less than a fortnight—formed the basis of one of the articles of impeachment. From the time when technology first made such tools as eavesdropping and wiretapping available, through the Nixon presidency, presidents have asserted their right to employ such devices in the furtherance of national security. Initially the Supreme Court distinguished, for Fourth Amendment challenges, surveillance techniques that required actual penetration of the premises from those that did not, but technology and sensible jurisprudence eventually erased that distinction. Of course, legal protection has never been extended to noncitizens of the United States whose conversations have been intercepted abroad.

In 1978, just one year after the Nixon interviews were broadcast, Congress enacted the Foreign Intelligence Surveillance Act (FISA), requiring the approval of special federal courts in situations where federal authorities sought to monitor phone calls between foreign nationals and U.S. citizens. During the first ten years of life under FISA, not a single monitoring application was rejected by the special court, and today, nineteen years later, fewer than a dozen have. Nonetheless, in December 2005, when *The New York Times* broke the story of the program's existence, George W. Bush's attorney general, Alberto Gonzales, claimed that the need for speed required circumvention of the courts. In January 2007, Gonzales informed the Congress that approval of the court would be sought in all future cases. No serious impeachment move based upon violation of the 1978 act or the constitutional rights afforded citizens was ever initiated. Nor did other actions such as the indefinite incarceration of terrorist suspects—sometimes without charges lodged against them—the denial of habeas corpus, and the asserted right to treat prisoners more roughly than the Geneva Convention permits give rise to any serious impeachment moves. It would seem that in certain respects, Nixon may well have had to confront a double standard. In his case, however, that standard was the product of self-inflicted wounds, including participation in a criminal cover-up, initiation of the "Saturday Night Massacre," and the bald assertion of discretionary power: "If the president does it, then it's not illegal."

Bewilderingly, despite the day's successes, the gloomy mood swallowing my staff had not abated. Marv Minoff and Jim Reston had circulated among the production crew at the end of the session and had heard little but praise for the way Nixon had "stuck it to the press." Zelnick's mind was more concerned with opportunities he thought had been missed than with openings seized. John was focused more on the disappointments of the morning than the extraordinarily high quotient of riveting material in the afternoon. And after months of round-the-clock work, they were all exhausted.

A rough-cut screening of the Vietnam program later that evening did little to allay the anxieties now eating away at my team. The cut was too long, too jagged, and still "too PBS." Clay Felker, who had just arrived in town, fell asleep. Though I felt strongly that the Vietnam sequence simply needed another pass, the raw material was all there—my colleagues worried that if we couldn't hold our own on Vietnam, Nixon would certainly eat us for breakfast on Watergate.

"But we have confronted him where it really matters," I argued. "Look at Chile. I certainly did not let him carry the day there."

"Red sandwiches and beef," said Bob. "I thought for a while that you were in a goddamn delicatessen. Don't kid yourself. He's set up perfectly now for kicking your tail from one end of Monarch Bay to the other. The only thing he has to watch out for is overconfidence. And that's not the position we should be in after more than two weeks on the set."

"That is not an accurate summary of the situation," I said, "any more than a goddamn delicatessen is an accurate summary of the Chile confrontation. You're forgetting the Kissinger material and the Mideast material. And you're forgetting that when we had to confront him, I did confront him. I confronted him on Chile, on Cambodia, yes on Vietnam, even on those crime statistics in the District of Columbia. Watergate will be like that, only ten times more so."

"Don't you know what you're up against?" said Zelnick. "This man is not only one of America's cagiest politicians, he's been a member of the bar for almost forty years. He's tried cases at trial, presided over committee hearings, argued before the Supreme Court of the United States. You've seen how formidable he is on matters he knows something about. Well, let me tell you, by next Wednesday, he will have committed every word on every Watergate tape to memory. He'll know every statute cold. He'll have rehearsed his answers. He'll call you every time you put a comma in the wrong place. He'll respond to your damaging references with dozens that support his interpretation of the facts. He'll put everything in a factual context that will take ten minutes to fabricate. You're in against a master, man, a master. Everything he wanted these interviews to accomplish for him will be on the line beginning next Wednesday. And he's a fighter."

"Yes, you can't back down with Richard Nixon," said John, echoing the sense of foreboding in Bob's words, "because he takes it as a sign of weakness. There's no mercy in the man, not as a warrior and not as an antagonist in this setting. He takes what he gets. You've got to stop him on the spot when he misrepresents the record and say, 'No, Mr. President, I know this better than you do, and I'm not going to let you rewrite the record.' Just look at this transcript and see how long his responses are. You have to declare your points. You have to stand with them. And you have got to destroy his points. Otherwise we will fail. You will fail."

"And right now," said Bob, "that's where we're heading."

Toward *failure*?

The gloomy attitude had gotten out of hand, and I was not

just annoyed, I was angry. What I'd heard was an absurdly lopsided account of the past few days, and indeed of the next few days. It was overkill. And it could be extremely destructive. So I decided to say so.

First, I reiterated the point that I thought should have been obvious without my having to mention it. Yes, I was daunted too. I realized how far we had to go. And time was indeed a commodity in terrifying—and legally limited—short supply. But that was no excuse for this sort of masochistic orgy.

Sure, the Vietnam edit wasn't working yet—but that was a problem that could and would be solved. If we had made mistakes, we still had the intelligence and the time to correct them. And though I agreed the first hour of domestic policy the day before had been Toilet City, the Huston plan was our "smoking gun." In the end, however, not only did I not believe their low estimate of our chances, I did not believe that deep down they did either.

"However, if there is anyone who really thinks that we are going to fail, it is better they leave this project now. It would just make the next few days too depressing." There was silence. The moment, it seemed, had passed. The taboo word had been uttered, faced, and rejected. We got back to work.

We divided the next five days in two: three days for separate effort, then two days for conferring together as needed. Zelnick spent his time working on a short but brutally effective paper, "Likely Nixon Detours," in which he set forth the possible excursions Nixon might lead us on in an effort to get up off the trail of his own criminal culpability. And he outlined the kinds of responses we would need to abort each such frolic and get us back in hot pursuit of the game.

Meanwhile, I spent my time poring over the White House transcripts, a few odd books, and excerpts of the Watergate trial testimony meticulously compiled by Reston. I was not yet at the point where I felt it necessary to develop a complete theory of the case against Nixon. That could wait until Monday, but I did begin to think in terms of what it was that I could specifically prove against the former president. I made little lists, checked them against the portions of the transcripts I had underlined, added to them, subtracted from them, changed the wording, put them aside, and began compiling new lists from scratch.

I was satisfied that I could prove that Nixon had ordered H. R. Haldeman on June 23, 1972, to order the CIA to request the FBI curtail its probe of the break-in.

I could prove that on January 8, 1973, Nixon had given Charles Colson favorable signals with respect to clemency for E. Howard Hunt.

I could prove that on February 13 and 14, again in discussion with Colson, he had staked his entire second term on the continued silence of the seven convicted defendants. The only problem would be "if one of the seven begins to talk."

I could prove that on March 13 he had ignored clear statements from John Dean regarding the criminal involvement of current members of his administration, including Gordon Strachan and Jeb Magruder.

I could prove that beginning March 20 he had leaned on Dean to write a phony document absolving White House personnel of culpability in the break-in and cover-up.

I could prove that during the March 21 through 22 marathon Watergate discussions he had been told of blackmail payments in process and later that they had been paid and had—at the very least—done nothing to turn them off.

I could prove that on March 21 he had coached Haldeman on how to commit what any reasonable person would call perjury without getting convicted.

I could prove that after Dean fell out of his good graces, he had turned Watergate matters over to Ehrlichman, who was himself deeply implicated in the cover-up conspiracy.

I could prove that between March 27 and April 14 he had attempted to get Magruder and Mitchell to come forward and take the rap, hoping the investigation would stop with them.

I could prove that through Haldeman and Ehrlichman he had offered suggestions of clemency to Mitchell, Magruder, and Dean in order to coax them into not implicating those even closer to him.

I could prove that he had shamelessly betrayed Henry Petersen's confidences by relaying his accounts of the investigation to Haldeman and Ehrlichman.

I could prove that between April 14 and 17 he had worked on "lines," "scenarios," and "drafts" with his two colleagues in order to explain their involvement in the earlier hush money payments.

I could prove that he had recommended no immunity for those suspected of criminal wrongdoing after Ehrlichman told him on April 17 that such action might be the best way to remove Dean's incentive to talk.

I could prove that on April 20 he recalled authorizing the blackmail payment to Hunt.

And I could prove that he had repeatedly and egregiously lied to the American people regarding his own state of knowledge at various times and his efforts to unravel the truth within his own White House.

What could I not prove? That list was almost as important as the first, though it was a good deal shorter.

I could not prove that he had had prior knowledge of the Watergate break-in.

I could not establish a motive for the original crime.

I could not establish a motive for his original involvement in the cover-up.

I could not prove that he had been aware of the activities of his private lawyer, Herbert Kalmbach, and Hunt during the summer and fall of 1972.

I could not prove that he had known of the blackmail payments before March 21, 1973.

I could not prove his personal involvement in the erasure of eighteen and a half minutes of his June 20, 1972, Watergate-related conversation with Haldeman.

I could not prove that he had personal knowledge of who had erased the tape, or, indeed, how the erasure had occurred.

I reviewed my list of "could not proves." With respect to some items the circumstantial evidence was strong. Would, for example, Ehrlichman and Haldeman have dared to enlist the president's personal attorney for the raising and transmittal of blackmail money without clearing it with Mr. Nixon himself? Would Haldeman have released $350,000 from Nixon's White House safe without Nixon's knowledge and consent? The gap in the tape had been created through five to nine manual erasures. Stephen Bull and Rose Mary Woods were the only ones with the possible motive to cause that erasure, except for the president himself. Would either of them have acted alone?

Still, it seemed the better part of wisdom to ignore these episodes or relegate them to a secondary line of questioning. Some would have to be covered just for the record. Others might simply be listed as additional questions concerning Nixon's conduct throughout the period. But I was convinced that our approach all along was sound: Stay with what we know we can prove. Keep within the essential cover-up period—June 1972 to April 1973.

By the time Monday morning arrived, I felt positively buoyant. I knew the transcripts inside out. I had John's questions, Zelnick's, Reston's, and mine. I had Bob's latest road maps—intended not as a text but as a guide—embracing as much supporting documentation as feasible for purposes of following up. They were already complemented by my own copious notes, as well as numerous excerpts from the exclusive Reston research material. What I had before me represented a true team effort.

One issue I wanted to work through as a group was what it would take to prove Nixon's guilt if this were a criminal conspiracy trial.

By Tuesday night, my team seemed to be brimming with confidence. Still, as we returned to our respective hotel rooms, each of us recognized that the fate of the entire project would likely be decided the next day.

It was after 1 A.M. when I responded to a faint knock at my door.

"Sorry to disturb you, David, but John asked me to deliver this to your room," said Libby, my PA and one of our tireless team. She handed me an envelope, wished me good luck the next day, and left.

Inside the envelope was a handwritten note John had penned on the hotel's stationery. It was, at one and the same time, the most inspiring and the most constructive letter I can remember receiving. The expressions of confidence in me, coming from a man who could not, indeed would not in principle, exhibit an ounce of false shmaltz to save his life, moved me more than I can say.

And his recital, which followed, of the basic components of our mutually agreed-upon strategy could not have been more economical or to the point:

Today you should review only the evidence that would have been brought up in a court of law: and you should not depart from that evidence . . .

It is not a conventional interview: you are exchanging interpretations of the known facts; you should talk almost as much as he does . . .

Most importantly, don't be tempted to put brief and "pointed" questions that elicit long and vague answers: when he paints a picture that you know to be false, respond by painting, at the same length he does, the alternative picture as you understand it . . .

Always keep firmly in mind that Watergate is a difficult subject for a mass audience to follow and at each stage consider that it is your responsibility to point out clearly to that audience the implication of any question, fact, event, statement or admission that you consider relevant . . .

Stay cool and firm, but be polite; only raise your voice if and when you are pushed to . . .

And finally, keep up the pressure at all times: You will win only if you can, so to speak, sprint the mile.

As he so often did, John had hit the nail on the head. I could not afford to relax *at all.* At no point tomorrow, even for a moment, were my interests and those of Richard Nixon going to coincide. My aim was to nail the basic truths about Nixon's involvement in the Watergate cover-up during an all too finite period of time. His aim was to filibuster, perhaps, or rather, to use his own words in our first interview, "to demolish." In a situation like that, anything I was going to get I was going to have to win.

On the ride to Monarch Bay that morning, I remember John adding something to the effect of "You know something, you have to stay physically in charge as well. There is something that you do sometimes . . . you did it on Cambodia and, although I wasn't there at the time, it seemed to me as though you did it with Chile . . . and you did it with Ian Smith when we were in Rhodesia."

I spent much of the drive receiving a primer—compliments of Bob Zelnick—on the law of conspiracy. "Now, let's see if I have this right," I said. "To be guilty of conspiracy, one must enter a scheme with at least one other person to commit an unlawful act and at least one member of the conspiracy must have performed at least one overt act in its furtherance, such as purchasing a gun with which to hold up a bank the conspirators intend to rob. Intent is an element of the crime, but motive is not. It is not a mitigation of the crime to say you intended to use your ill-gotten gains for benign purposes . . . or to mislead investigators who might cause you political embarrassment.

"We've talked a lot about the conspiracy to obstruct justice. Has anyone here read the actual statute?"

"Just a minute," said Jim. "I think there's a copy of it here in the Judiciary Committee volumes."

He quickly found the place and handed me the book. I began to read aloud: " 'Whoever corruptly endeavors to prevent, obstruct or impede the administration of justice . . .' Bob, that sounds pretty wide."

"I recall John Dean telling the Ervin Committee that the definition of the crime was coextensive with the capacity of man to devise ways to prevent, obstruct, or impede the administration of justice. In a sense, then, it means everything because it means nothing," Bob said.

"A corrupt endeavor to prevent, obstruct, or impede," I repeated. "Thank you, Jim. It may come in handy."

THE INTERVIEWS: DAY 8

April 13, 1977

Now the time had come to discuss Watergate. We arrived early, as I wanted to thank the audio men, technical directors, crews, makeup artists, and all others involved in the production. Their work would soon become the raw material of history. The series, particularly Nixon's words, voice, facial expression, and body language, would be studied by lawyers, presidential scholars, journalists, and others. We were tackling our most sensitive topic yet, and everyone wanted to know what Nixon would say. For that reason, maintaining security was essential. To date, the crew had been magnificent.

"Gentlemen, you've all been superb," I began. "Every reporter in the country wants to know what the former president has to say about Watergate. I must ask you not to tell them. Don't even tell your wives or the women you love. In fact, please don't tell either of them. Thank you."

Nixon arrived at his customary time, ten after ten. He greeted a cluster of local residents, exuding confidence. Then a car pulled up to the curb and from it stepped Diane Sawyer and Frank Gannon. They had been working on Nixon's memoirs and had not attended the earlier tapings. As the Nixon camp gathered on the curb, my own team eyed them warily.

"They're all here," said Birt.

"Yes," said Reston. "All the president's men."

I have many times tried to recapture my feelings as the countdown and cue came for me to begin. Long ago I settled on "euphoria." It was almost too much to bear. I started to speak.

"Mr. President, to try and review your account of Watergate in one program is a daunting task. But we'll press first of all through the sort of factual record and the sequence of events as concisely as we can to begin with. But just one brief preliminary question. Reviewing now your conduct over the whole of the Watergate period, with additional perspective now, three years out of office and so on, do you feel that you ever obstructed justice or were part of a conspiracy to obstruct justice?"

No goal with the first kick. Nixon wanted to go through the record. Then he would explain his motives. "I will give you my evaluation as to whether those actions or anything I said for that matter amounted to what you have called obstruction of justice."

I had no problem with that and launched an inquiry into events following the June 17, 1972, break-in at the Democratic National Committee headquarters. By the twentieth, Haldeman knew enough to order Gordon Strachan to make sure his White House files were "clean" and Ehrlichman knew enough to order Howard Hunt to leave the country, an order he soon rescinded.

Haldeman met with Nixon on the morning of June 20, a session memorialized by an eighteen-and-a-half-minute gap in the White House tapes, but also some handwritten Haldeman notes indicating that the president wanted a "PR offensive" and a check of the Executive Office Building (EOB) for Democratic bugs. "Haldeman's notes are the only recollection I have of what he told me," said Nixon. It was a statement neither of us believed but that accurately portrayed the sort of narrow defendant posture he would assume for most of the first session. He volunteered to debunk the "outrageous" suggestion that he had been responsible for the gap, another old defendant's ploy: attack what is not alleged.

I next hit Nixon with questions about the June 20 conversation with Colson that Reston had unearthed, and Nixon never knew we had done so. I recalled for him the talk of "pulling it all together," of the arrested men being "pretty hard-line guys," and Nixon's incautious remark "If we didn't know better, we would have thought the whole thing was deliberately botched."

A momentary shadow crossed his face, but Nixon quickly dismissed the new information, then introduced his leitmotif—an argument he would lean on whenever the going got particularly tough. "Let me say as far as what my motive was concerned," began Nixon. "My motive was in everything I was saying or certainly thinking at the time, ah . . . ah . . . was not, ah . . . to try and cover up a criminal action but . . . to be sure that as far as any slip-over or should I say, slop-over, I think would be a better word, any slop-over in any way that would, ah . . . damage innocent people or blow it into political proportions." The defense played directly into our hands and our understanding of the law. "In other words," he said, "we were trying to politically contain it." That had been his motive, he insisted, in what had come to be called the "smoking gun" tape.

"So you invented the CIA thing that day as a cover," I charged.

Nixon recoiled at the words.

"No. Now let's use the word *cover-up* in the sense that it has—should be used. If a cover-up is for the purpose of covering up criminal activities, it is illegal. If, however, a 'cover-up,' as you have called it, is for a motive that is not criminal, that is something else again. And my motive was not criminal." He hadn't known that Mitchell or anyone else was involved. He had worried instead that the FBI investigation could spring leaks, which would lead to the matter being "blown out of proportion." What's more, it would be good if the CIA could help protect Hunt, one of its own.

This was one of the historic Nixon tendencies. When he dissembled, he dissembled in all areas at once, partially and totally. Inconsistencies bothered him not at all. In response to a single question, he had denied the existence of a cover-up, admitted involvement in a cover-up, defined the term in a way no one knowledgeable of the law ever would, denied knowing anyone else was criminally involved, and then conceded his effort to help Hunt, whom he knew to be criminally involved, along with Liddy.

One of Nixon's more fascinating inconsistencies was his insistence that until he heard the final section of the June 23 tape on August 5 a year later, just days before his resignation, "I thought and often stated, ah . . . that the purpose of getting the CIA in was primarily or solely for national security reasons being that, the reasons being that the FBI investigation, ah . . . might, ah . . . infringe upon investigations or activities of the CIA which they wanted to be kept covert." Was Nixon really expecting our audience to believe that he had misremembered his own motivations? Indeed he was. According to Nixon, his busy schedule had obscured such recollections.

We tussled over this until he introduced his July 6 meeting with FBI Director-Designate L. Patrick Gray, who warned him "that there are some people around you who are mortally wounding you or would . . . might mortally wound you because they're trying to restrict this investigation." Nixon asked Gray if he had talked to Walters about the matter, and Gray said he had and that Walters was in agreement. At that point, recalled Nixon, "I said, 'Pat, you go right ahead with your investigation.' He has so testified. And he did go ahead with the investigation."

There were two problems with Nixon's reliance on that instruction to purge himself of conspiracy in the cover-up. First, telling Gray to get the facts, while simultaneously working with his White House subordinates to keep the facts from coming to light, compounded the crime. Second, the conspiracy had already been busy at work at least from June 23.

"Obstruction of justice is obstruction of justice if it's for a minute or five minutes . . . much less for the period of June twenty-third to July the fifth." I said. "It's obstruction of justice for however long a period, isn't it? And it's no defense to say that the plan failed, that the CIA didn't go along with it."

These were telling points, and Nixon tried to fend them off by questioning my knowledge of the obstruction of justice statute.

"Now just a moment," he began. "You're again making the case, which of course is your responsibility as the attorney for the prosecution. Let me make the case as it should be made—even were I not the one who was involved—for the defense. The case for the defense here is this: You use the term obstruction of justice. You perhaps have not read the statute with regard to respect . . . ah . . . ah . . . obstruction of justice."

"Well, I have," I interjected, sorely tempted to tell him that I had inspected it only minutes before coming onto the set. But even my more modest claim knocked him off stride to the point where he had trouble articulating his words.

"Obstructed—well, oh, I'm sorry, of course you probably have read it, but possibly you might have missed it because when I read it, many years ago in, ah . . . perhaps when I was studying law . . . if, although the statute didn't even exist then, because it's a relatively new statute, as you know." (At this moment, the laughter in the production trailer could have registered on the Richter scale.)

"Ah, but in any event," Nixon continued, "when I read it, even in recent time, I was not familiar with all of the implications of it. The statute doesn't require just an act. . . . The statute has the specific provision . . . one must corruptly impede a judicial—"

"Well," I interrupted, "a corrupt endeavor is enough."

"Conduct . . . all right . . . we'll . . . a conduct endeavor . . . corrupt intent, and that gets to the point of motive. One must have a corrupt motive. Now, I did not have a corrupt motive."

I also reminded him that the criteria of conspiracy law, passed during his administration, had been fulfilled when he, Haldeman, and Ehrlichman had conspired to obstruct justice. "And," I continued, "motive can be helpful when intent is not clear. Your intent is absolutely clear. It's stated again: 'Stop this investigation here.' The foreseeable, inevitable consequence, if you'd been successful, would have been that Hunt and Liddy would not have been brought to justice." Clearly the thrust of my questioning linked Nixon with his closest administration colleagues—Haldeman, Ehrlichman, and Mitchell, all of whom had been found guilty by the time I interrogated Nixon.

On the subject of hush money, Nixon insisted that, while he had not been informed, "if I had been informed the money was being raised for humanitarian purposes to help these people with their defense, I would certainly have approved it."

"Right," I said, "and if you'd been told that they were saying that it was for humanitarian reasons but it was being delivered on the tops of phone booths, with aliases, and at airports and with people with gloves on . . . would you have believed that it was for humanitarian reasons? That's not normally the way that lawyers' fees are delivered, is it?" Nixon conceded that "of course, I would have had a suspicion about it."

Nixon had insisted that it was not until March 20, 1973, that he learned the money had been used to buy the witnesses' silence. But I noted earlier conversations in which Ehrlichman had told him the money was needed to keep defendants from going "off the reservation." And Haldeman had told him it was because the defendants were "gonna blow." "Yeah," said Nixon, "well, it has more meaning, but let's understand what the word 'blowing' means too."

I was incredulous. "But when he said that, you said, 'Come on, we've got to think of a better story than that.' And you still haven't."

Nixon continued to maintain that only on March 21 had he learned most details of the cover-up. So I confronted him with an excerpt from a February 14 conversation with Colson—another Reston tape Nixon did not know we had. "The cover-up is the main ingredient. That's where we've got to cut our losses; my losses are to be cut. The president's losses got to be cut on the cover-up deal."

Here Nixon gave one of those awkward waves of his arm that reminded me of no one so much as the late Jack Benny and again suggested he was concerned mainly about the defendants going to the press. He had a tougher time with a statement to Colson in a February 13 conversation, yet another tape he had not known we had: "When I'm speaking about Watergate, though, that's the whole point of the election. This tremendous investigation rests, unless one of the seven begins to talk. That's the problem." Clearly the integrity of Nixon's second term *did* rest on assurances by the president and top law enforcement officers that Watergate was being fully and fairly investigated. The mandate Nixon sought and achieved rested on them, the reorganization of government, the pursuit of détente, the new conservatism—it was all there before him, "unless one of the seven begins to talk." Once again Nixon insisted that his fear was that the men who worked in covert activities would breach security by talking.

We came to the pivotal conversation with Dean of March 21. Not only was it clear that the president wanted hush money paid, but the conversation proved to me beyond a reasonable doubt that it had been Richard Nixon, not Dean, not even Haldeman, who urged that the cover-up continue. Dean had warned the president in the starkest possible terms that Watergate had become a "cancer on the presidency" that, unless cut out, could destroy the presidency. He had warned that several people, including himself, would likely go to jail.

Whatever his underlying motives, however "blind" his past ambition had been, Dean was now providing the president with all he needed to know to make an honest decision about handling the crisis. The army of Nixon defenders who pounced on Dean for his Ervin Committee "treason" could not point to a single change that would have put the president in a position to make a better decision.

It was Nixon who repeatedly underlined the desirability of meeting Hunt's pressing financial demands, who repeatedly returned to the subject of the money and the need to pay it. And if Nixon could later find, here and there, a saving reference to the difficulty of early clemency, he would find the record barren of any suggestion that he had so much as lifted a finger to bring the course of continuing criminality to a halt.

Once again Nixon began by suggesting it was not Watergate but other matters Hunt had performed for Ehrlichman that concerned him, matters such as the Daniel Ellsberg problem. This hardly amounted to a defense, as the break-in into the office of

Ellsberg's psychiatrist had been no less illegal than Watergate. Other Hunt matters included getting dirt on Ted Kennedy, forging cables to suggest that JFK had approved of the assassination of Ngo Dinh Diem, and spiriting the ITT lobbyist Dita Beard out of town when she began to cause political embarrassment. But rather than continue playing word games with the former president, I decided to confront him with a list of his own most damning quotes to Dean, something I had been working on over the weekend.

> "You could get a million dollars, and you could get it in cash. I know where it could be gotten."
>
> "Your major guy to keep under control is Hunt."
>
> "Don't you have to handle Hunt's financial situation?"
>
> "Let me put it frankly: I wonder if that doesn't have to be continued?"
>
> "Get the million bucks, it would seem to me that would be worthwhile."
>
> "Don't you agree that you'd better get the Hunt thing?"
>
> "That's worth it, and that's buying time."
>
> "We should buy time on that, as I pointed out to John."
>
> "Hunt has at least got to know this before he's sentenced."
>
> "First, you've got the Hunt problem. That ought to be handled."
>
> "The money can be provided. Mitchell could provide the way to deliver it. That could be done. See what I mean?"
>
> "But let's come back to the money. ["They were off on something else there," I commented. "Desperate to get away from the money; bored to death with the continual references to the money."] A million dollars and so forth and so on. Let me say that I think you could get that in cash."

"That's why your immediate thing, you've got no choice with Hunt but the hundred and twenty or whatever it is. Right?"

"Would you agree that this is a buy-time thing? You'd better damn well get that done, but fast."

"Now who's going to talk to him? Colson?"

"We have no choice."

I had gathered momentum as I went along. Nixon remained guarded. His countenance was placid through the first several items, then his lips quivered. His eyelids fluttered like the wings of a moth shot through with electric current. His head lurched backward with each new item. He was a man in pain, a man on the ropes, but not yet a man ready to concede defeat.

"Let me stop you right there, right there!" He accused me of reading the quotes "out of context, out of order." I felt the cumulative impact was worth this bit of poetic license.

Nixon kept insisting that his remark late in the conversation "You never have any choice with Hunt, because it finally comes down to clemency," reinforced by his statement "You can't provide clemency," amounted to a rejection of hush money payments. I wanted to be fair, but I could never make that link. Clemency was a long-term proposition. A better solution would have to be found. The $120,000, though, was a "buy-time thing." There was nothing mutually exclusive or even mildly inconsistent about ratifying the blackmail payment and declining to go ahead with clemency. An even more damning fact: "Here's Dean, talking about this hush money for Hunt; talking about blackmail and all of that. I would say that you endorsed or ratified it. But leave that to one side—"

NIXON: I didn't endorse or ratify it.

FROST: Why didn't you stop it?

Again the question jolted Nixon. He tried to collect himself. "Because at the point, I had nothing to—no knowledge of the fact that it was going to be paid."

It was time to wrap our morning session. Birt and Zelnick bolted from the trailer. Zelnick met me at the door of my room.

"David, it was super! First-rate! Sensational!"

We embraced and looked around for John. He was engaged with the Nixon staff outside their monitoring room. Brennan and Khachigian had stopped him as he followed Bob. "What a mistake," said Brennan. "What a mistake."

"The president of the United States made himself look like a criminal defendant. With David as prosecutor," Khachigian agreed.

"We didn't want him to go that route," said Brennan.

"But this was one subject that we simply could not discuss with him. It was just too personal," said Khachigian.

"That's right," said Diane Sawyer, joining in the conversation. "He hasn't written the Watergate part of his book yet. So none of us knew what he was going to say."

THE INTERVIEWS: DAY 9

April 15, 1977

The fog had lifted. The cloud of gloom was only a memory. It had evaporated once and for all into one and a half hours of brilliant morning sunshine. Marv's face took on the look of a man ten years younger. Jim, a calm presence even during moments of intense dispute, seemed now like a man who had achieved nirvana, his countenance a portrait of total enlightenment and peace. Bob, who gave ground grudgingly when he thought things were not going well, now with equal passion defied one and all to suggest so much as a single imperfection in the morning's proceedings. There were instant replays of every pivotal exchange.

But now it was time to look ahead. John had a particular concern. "I don't know what Nixon will come up with on Friday. Jack and Ken seemed so genuinely disappointed when I talked to them yesterday that I feel sure that they would do everything in their power to reach Mr. Nixon and persuade him to take a different line. I don't know what it will be, but I think we will have to be ready for it."

Bob expected no strategic adjustment. "Contrition in any meaningful sense is alien to Nixon's personality. He is psychologically incapable of it. We are going to face the same stonewall tomorrow we faced on Wednesday."

At the end of our preparation session, I asked the team, "Assuming what Bob said earlier about the stonewall is right, and taking John's point, do you think that at the end we ought to invite him expressly to retract the hard-line approach and go the other route? Do you think the occasion calls for it?"

"Absolutely," said John. "Whether or not you agree with it—and I don't, by and large—I feel he's presented a coherent view of himself and his administration except in these abuse areas. But as long as he remains rigid there, the rest of his record will never be taken and debated seriously."

"Bob," I said. "You've been so adept to anticipating Nixon's responses all along, how do you think he'll reply to an emotive challenge like this?"

"His face will contort," said Bob, "His eyes will glisten. His voice will break. His head will nod gently and sadly. With the weight of history on his every word, he will say . . . 'Screw you.' "

Nixon arrived on the set seventeen minutes late—his first missed "deadline."

I told him we would pick up where we had left off the previous day. He was advising Dean and Haldeman that "perjury is a tough rap to prove." Then his advice became even more specific: "Just be damn sure you say, 'I don't remember . . . I can't recall.' "

"Is that the sort of conversation that should have been going on in the Oval Office, do you think?"

Instead of bristling at the new allegation, Nixon seemed anxious to respond to this new and serious charge.

"I think that kind of advice is proper advice for one who was at that time . . . beginning to put myself in the position of an attorney for the defense—something which I wish I hadn't had the responsibility, felt I had the responsibility to do. But I would like the opportunity when the question arises to tell you why I felt as deeply as I did at that point."

I turned to the Dean report. In an August 15, 1973, statement, he had explained its origins to the American people as follows: "If anyone at the White House or high up in my campaign had been involved in wrongdoing of any kind, I wanted the White House to take the lead in making that known. On March 21, I instructed Dean to write a complete report of all that he knew on the entire Watergate matter."

I compared that with the true record. On March 17, the president had asked Dean for a "self-serving goddamned statement" denying the culpability of the principal figures. When Dean had told the president that the original Liddy plan had involved bugging, Nixon had told him to omit that fact from his document. "On March twenty-first," I told Nixon, "after his revelations to you, you said, 'Understand, I don't want to get all that specific.' "

Nixon was still playing defense. He reminded me that Dean had not mentioned Magruder's admitted involvement in planning the initial Watergate break-in. I reminded Nixon of just how much information had in fact been conveyed by Dean on March 21. The central figures—Magruder, Kalmbach, Haldeman, Ehrlichman, Mitchell, even Dean himself—had all been named and their transgressions listed. "And therefore, when you say, 'This person isn't involved and that person isn't involved,' you knew they *were* involved." Then a stunning breakthrough. Nixon, finally, began to crumble.

Nixon: "I would think you would have found that statement . . . Let's get an impression of the whole story. Let the bad come out . . . there's plenty of bad. I'm not proud of this period. Ah . . . I didn't handle it well. I messed up."

Here he recalled Mayor Fiorello La Guardia's famous remark "When I make a mistake, it's a beaut."

"Well," he continued, "I must say, mine wasn't a beaut—it was a disaster. Ah . . . and I recognize that it was a mistake, I made plenty of them. Ha, but . . . ah . . . I also insist that as far as my mistakes were concerned, ah . . . they were mistakes, frankly, of the head and they weren't mistakes of the heart. They were not mistakes that had what I call an improper, illegal motive, ah . . . in terms of obstructing justice. Ah . . . that's all I'm trying to say."

This, I suspect, was the statement Nixon had come prepared to make. A simple declaration that he had made disastrous errors of the mind but not the heart, and that reading the totality of his words in context, one could see that he had been involved in no criminal conspiracy to obstruct justice.

I continued to pepper Nixon with excerpts from conversations with Ehrlichman and Haldeman, firmly establishing the Dean report as an instrumentality of the cover-up, rather than as its exposition. He gave ground foot by foot, yard by yard.

"I still don't know why you didn't pick up the phone and tell the cops," I said. "I still don't know, when you found about the things that Haldeman and Ehrlichman had done, that there is no evidence anywhere of a rebuke, but only of scenarios and excuses, et cetera. Nowhere do you say, 'We must get this information direct to whosoever it is—the head of the Justice Department criminal investigation or whatever.' And nowhere do you say to Haldeman and Ehrlichman, 'This is disgraceful conduct'—and Haldeman admits a lot of it the next day, so you're not relying on Dean—'you're fired.' "

Nixon's body seemed to go limp. It was clear he was searching for ways to say new and different things. Up to this point the confrontation had been Nixon versus Frost. Now the former president seemed to be battling within himself. Nixon asked for time to address the question. But then he began wandering from Vietnam to his ambitions for the second term and his lack of knowledge about the law relating to cover-ups.

I was determined now to interrupt, but then I deferred when he again started talking about his special relationship with Haldeman and Ehrlichman.

"But when it came to March twenty-first," I reminded him, "and the revelation . . . ah . . . Haldeman and Ehrlichman soon made it very clear to you that these payments, for instance, were not in fact innocent payments for humanitarian reasons . . . I mean, Haldeman said the defendants 'might blow.' Ehrlichman said they were there to keep him [Hunt] 'on the reservation.' " I compared instructing Ehrlichman to conduct an independent inquiry with asking Al Capone for an independent investigation of organized crime in Chicago.

Nixon digressed to the Sherman Adams affair, an incident from the Eisenhower administration when Ike had been forced to let go of his chief policy adviser, who had accepted gifts from a businessman named Bernard Goldfine. Ike had designated his vice president—Richard Nixon—to handle the unpleasant chore of notifying Adams. With Haldeman and Ehrlichman, Nixon had no appropriate surrogates. He alone had to tell them they were through.

With Haldeman and Ehrlichman, Nixon also had to balance the value of keeping their good counsel close at hand versus that of purging the administration of those stained by Watergate. Nixon considered pardons for those involved, somehow combining those with assertions of the doctrine of executive privilege.

In the end, Nixon issued no pardons. In my view, he was simply too proud. He wanted to have it both ways, appearing publicly as a law-and-order president, devoted to unraveling the cover-up, yet privately conspiring to make the cover-up work. Nixon's indecisiveness doubled his problems. Given a choice between one form of damage control and another, the president—on three pivotal occasions—chose both.

Nixon and his two aides repaired to Camp David. Haldeman was the first to learn his fate, telling Nixon, "I disagree with your decision totally; I think you're going to live to regret it." Haldeman was bitter, feeling he had done nothing wrong. Nixon took Ehrlichman out to the porch. "I said, 'You know, John, when I went to bed last night . . . I hoped, I almost prayed, I wouldn't wake up this morning.' Well, it was an emotional moment; I think there were tears in our eyes, both of us."

Nixon quoted William Gladstone as advising those who wished to be prime minister that the job required one to be "a good butcher."

"Well, I think the great story, as far as a summary of Watergate is concerned," Nixon said, "I did some of the big things rather well, I screwed up terribly on what was a little thing and became a big thing. But I will have to admit I was not a good butcher."

I asked Nixon if it was possible for him to "go a little further." He had spoken in terms of loyalty to his friends and his perceived need to serve as defense counsel for them, but he had also told the nation that his sole concern during the period March 21 to April 30 was to get the truth out. "But now you're telling us your innermost feelings at that time, and I've indicated some of my doubts, and I've got others, about the speed with which the truth came out."

He was still not ready to go further. I began looking for an excuse to break. In the corner of my eye I thought I saw Jack Brennan holding up a piece of paper saying, "Let us talk." I told Nixon we needed time to change tapes. I started for our own monitor room, but Brennan was waiting in the hall. His face was flushed. He began to talk in a jumble of words. I heard only isolated phrases. "Critical moment in his life." "Can't cross-examine him." "Know he'll go further." "What do you want?"

On the floor lay his piece of paper. It did not say "Let us talk," it said "Let him talk." Khachigian came rushing out of Nixon's room, whispered something to Brennan, and rushed back in. Then Birt arrived from the production trailer.

"What is it you're trying to say, Jack?"

"David has got to quit playing the prosecutor," Jack said. "This is an important moment in the president's life. He'll go further than mistakes and misjudgments. He wants to make a full accounting. But you've got to let him do it in his own way."

"What do you mean by a full accounting?" asked Birt. "That he was guilty of a crime?"

"I don't know if he'll say that."

"That he committed impeachable offenses?"

"I don't know if he'll say that either."

"Then David's cross-examination will resume."

"Just a minute. Let me talk to him."

Birt turned to me. "That was exceptional, David. But we can't relent now."

"No, we can't. On the one hand, Jack's right when he says Nixon won't go much further under adversary pressure. On the other hand, I have to dispute or at least disclaim any categorization of his conduct which doesn't reflect ours."

"Don't change a thing," said John.

Amid all the emotional turmoil, Brennan returned from the Nixon room.

"He knows he has to go further," he said. "I don't know what he'll say, and I'm not sure he does. But ask him. Just ask him. He's got more to volunteer."

"Look, Jack, we can't plea-bargain with you," said John.

"If he's got something to say," I said, "we'll give him every opportunity to say it on his own, but if he falls short, we'll have to come back at him."

"The interrogation will have to restart. That's all we can tell you now," said John.

"I'll go and tell him," said Brennan, "and if it doesn't happen now, we can always try again on Monday." John put out his hand and stopped Brennan for a moment. "No, Jack," he said with intense earnestness. "Don't let him feel that even for a second. Believe me, if he doesn't do it now, having come this far, he'll never do it on Monday."

I have often read of "electricity" in the air. Of a "highly charged" atmosphere. But I never expect to experience it again quite as I did as Nixon and I walked back on to the set. Everybody felt it. All over the house. John felt it on his way back to the trailer. He stopped, turned, and came back again. He walked across the set to the side of my chair, leaned over, and whispered in my ear, "It is terribly easy for all of us to get caught up in the emotions of the moment. But what happens now and what he says now, and what you say now, will be pored over by historians. That's the perspective to try and keep."

The red light on the cameras went on. I tried to recapture the mood of our earlier conversation.

"To come back to where we were just now, Mr. President . . . because this is a difficult program for you and a difficult program for me. We were talking about the period . . . March the twenty-first and April the thirtieth. And you were talking about your emotions as you had to bid farewell to Haldeman and Ehrlichman. And talking about the mistakes that you made and so on in doing that . . . you've talked about the mistakes . . . we're at an extraordinary moment in a way. Would you do what the American people yearn to hear? Not because they yearn to hear it, but just to tell all, to level and so on. Would you go further than the 'mistakes'? You've explained how you got caught up in this thing. You've explained your motives. I don't want to quibble about any of that. But just coming to the sheer substance—would you go further than 'mistakes'? The word that seems not enough for people to understand."

"Well, what word would you express?"

This was the most heart-stopping moment I have ever had in an interview. Richard Nixon was probably more vulnerable than he would ever be again. And he was putting the question back to me. It was a moment I had to seize. Unless I was able to frame, with precision, what it was we wanted to hear from him, the moment would be lost, never to be recaptured. As a symbolic gesture, I took the clipboard with my notes on and tossed it onto the floor beside my chair, trying to indicate that whatever I was about to say was not some prepared ploy. It could never have been prepared, of course, because none of us had been anticipating a moment like this.

"Let me say that my concern is now not to—which is why I chucked the clipboard away—not to be legalistic or anything, about obstructions of justice and so on, and things we've discussed so far and so on . . . I think there are three things—since you asked me that heart-stopping question—I would like to hear you say, and the American people would like to hear you say. One is, 'There was probably more than mistakes . . . there was wrongdoing.' Whether it was a crime or not—yes, it may have been a crime too. Secondly, 'I did'—and I'm saying this without questioning the motives, right—'I did abuse the power I had as president or not fulfil the totality, the oath of office.' That's the second thing. And thirdly, 'I put the American people through two years of agony, and I apologize for that.'

"And I say that—you've expressed your motives—I think those are the categories. And I know how difficult it is for anyone, and most of all you, but I think that the people need to hear, and I think unless you say it, you're going to be haunted for the rest of your life."

Over the next twenty minutes, Nixon addressed all three of these points, but my recital seemed momentarily to drive all the air from his lungs. Then he began slowly, circuitously. No, it had not been a good time for the country. He had made mistakes—"horrendous ones . . . ones that were not worthy for a president . . . ones that did not meet the standards of excellence that I had always dreamed of as a young boy."

He had considered resigning on the day he announced the resignations of Haldeman and Ehrlichman. Along with the bad, he "owed it to history" to remind Americans of some of his great accomplishments—the second and third summits, resolution of the crisis in the Mideast, the continuing processes of détente and normalization with the two Communist powers.

He returned to the point. He hadn't just made mistakes in this period. Some that he regretted most deeply involved "the statements that I made afterwards" about his claimed efforts to unravel the cover-up.

"I would say that the statements I made afterwards, on the big issues, true: that I was not involved in the matters that I had spoken to . . . not involved in the break-in, that I did not engage in and participate in or approve the payment or the authorization of clemency, which, of course, were the essential elements of the cover-up."

But he had been in a "five-front war" with a partisan media, a partisan Ervin Committee, a partisan special prosecutor's staff, and a partisan House Judiciary Committee staff. "Now, under all these circumstances, my reactions in some of the statements and press conferences and so forth after that, I want to say right here and now, I said things that were not true. Most of them were fundamentally true on the big issues, but without going so far as I should have gone, and saying, perhaps, that I had considered other things but not done them."

FROST: Well, you mean—

NIXON: And for all those things, I have a very deep regret.

FROST: You got caught up in—

NIXON: Yeah.

FROST: —and then it snowballed.

"It snowballed," said Nixon. But he quickly repeated his claim that "on the essential issues, I leveled with the American people and told the truth."

Yet in the face of continuing attacks, his credibility began to go down at home, and it went down abroad. "By the time I resigned, I was crippled. I was crippled even before that."

He would take the blame for that. He was not blaming anyone else, certainly not Mitchell, Ehrlichman, Haldeman, and the rest. They had all suffered enough.

"I'm simply saying to you that as far as I'm concerned, I not only regret it, I indicated my own beliefs in this matter when I resigned. People didn't think it was enough to admit mistakes. Fine. If they want me to get down and grovel on the floor, no. Never. Because I don't believe I should."

Those words were spoken with conviction, even defiance. But he quickly came back to his more wistful, conciliatory tone. He was, again, not blaming anyone else, not the CIA and not his Democratic and Republican foes, the so-called impeachment lobby. He would reject the claims of those who called him a victim of a coup or a conspiracy.

"I brought myself down. I gave them a sword. And they stuck it in. And they twisted it with relish. And I guess if I'd been in their position, I'd have done the same thing."

Nixon had moved a tremendous distance. Could I get him to go even further? In addition to making these untrue statements, could he say with conviction "that you did do some covering up? We're not talking legalistically now, I just want the facts . . . that there were a series of times, maybe overwhelmed by your loyalties or whatever else, but as you look back at the record, you behaved partially protecting your friends—or maybe yourself—and that in fact you were, to put it at its most simple, a part of a cover-up at times?"

"I did not in the first place commit a—the crime of obstruction of justice—because I did not have the motive required for the commission of that crime." I reminded Nixon that we would continue to disagree on that. He responded that the matter was one for lawyers to argue and that the House of Representatives had ruled overwhelmingly that he *had*. But, rather than face a long Senate trial, which the country could not afford, "I have impeached myself. That speaks for itself."

FROST: How do you mean, "I have impeached myself"?
NIXON: By resigning, that was a voluntary impeachment.

During the critical post–March 21 period, "I was in a very different position. And during that period, I will admit that I started acting as a lawyer for the defense.

"I will admit that acting as a lawyer for their defense, I was not prosecuting the case."

And during that period, "As the one with the chief responsibility for seeing that the laws of the United States are enforced, I did not meet that responsibility.

"And, to the extent that I did not meet that responsibility, to the extent that within the law, and in some cases going right to the edge of the law, in trying to advise Ehrlichman and Haldeman and all the rest as to how best to present their cases, because I thought they were legally innocent, that I came to the edge.

"And under the circumstances, I would have to say that a reasonable person could call that a cover-up."

Nixon drifted slightly following this stunning series of admissions, but he had one more message to deliver to the American people. He got into it by recounting his farewell dinner at the White House with his closest congressional supporters. As he had risen to leave for the Oval Office and his final address as president, he had turned and told his friends, "I'm sorry, I just hope I haven't let you down."

"Well, when I said, 'I just hope I haven't let you down,' that said it all.

"I had.

"I let down my friends.

"I let down the country.

"I let down our system of government and the dreams of all those young people that ought to get into government but think it's all too corrupt and the rest.

"Most of all, what I fear the greatest—not that I don't hope and pray that President Carter will be able to make progress in his peace initiatives—I let down an opportunity that I would have had for two and a half more years to proceed on great projects and programs for building a lasting peace, which has been my dream, as you know from our first interview in 1968, before I had any thought I might even win that year. (I didn't tell you I didn't think I might win, but I wasn't sure.)

"Yep, I . . . I, I let the American people down. And I have to carry that burden with me for the rest of my life.

"My political life is over.

"I will never yet, and never again, have an opportunity to serve in any official position. Maybe I can give a little advice from time to time.

"I can only say that in answer to your questions, that while technically I did not commit a crime, an impeachable offense . . . these are legalisms.

"As far as the handling of this matter is concerned, it was so botched up.

"I made so many bad judgments, the worst ones, mistakes of the heart rather than the head, as I pointed out.

"But, let me say, a man in that top judge . . . top job, he's gotta have heart.

"But his head must always rule his heart."

Nixon's mea culpa had gone so much further than we had originally expected or even hoped. I was moved—awed—by the experience we had just shared. I suggested to Nixon that whatever burdens he had been shouldering would be lighter now.

"I doubt it," Nixon replied. In his view, his mortal enemies among the media, liberal Democrats, and academics would never surrender their hatred for him.

Nixon said good-bye and started to leave. But he was ambushed by the two staffs, which had already converged in the living room. We were both surrounded and congratulated for one of the most extraordinary moments any of us could ever recall.

"Do you think they'll accept what he said as satisfactory?" Khachigian asked after Nixon had left.

"I would certainly hope so," said Zelnick. "The president was as honest today as God has given him the capacity to be honest."

"It's funny," Diane Sawyer mused sadly, "you people are journalists, and good ones, and you probably learned more about Richard Nixon than any other outsiders in the world; sometimes I think you know him better than we do. But I think we know more about your colleagues than you do. Just watch. They're going to see your show. And they're going to tear him to shreds."

"Wanna bet?" asked Zelnick, extending his hand.

"Why not?" Diane replied, clasping it. "We've got nothing to lose."

THE INTERVIEWS: DAYS 10–12

April 18, 20, and 22, 1977

We had three more sessions scheduled with Nixon. One was devoted mainly to separate individual interviews we had agreed to do for our co-production partners, the BBC, TF1 in France, RAI in Italy, and Channel 9 in Australia. Each country sent one of its leading current affairs producers to work with us on a one-hour interview that would be for broadcast exclusively in their respective territories.

Then it was back to the main theme—the international interviews for Frost/Nixon. Among the most interesting items on our agenda was the break-in into the Los Angeles office of Daniel Ellsberg's psychiatrist, Dr. Lewis Fielding. Nixon came prepared to talk about national security leaks. He said there had been forty of them in 1969 and seventy-one in 1970. But by mid-1971 alone, the number had shot up to eighty-two. So the mandate of the White House "plumbers' group" formed by the president was to deal quickly with the greater problem.

Nixon was not modest in discussing Ellsberg. He said he had instructed Bud Krogh of the plumbers to determine what classified documents Ellsberg had access to besides the now-famous Pentagon Papers. Almost as an afterthought Nixon added a second motive, that of publicly discrediting Ellsberg. It was Kissinger, Nixon said, who became the great administration advocate for an anti-Ellsberg campaign. Leaks like Ellsberg's, said Kissinger, inhibited his ability to conduct diplomacy. Already there were cables from Canada, Australia, and Romania expressing concern about the security of their correspondence with the United States. The order to discredit Ellsberg had come from Nixon himself, "as I said, not Ellsberg as a man, not for the purpose of getting him convicted at a trial, but that would discourage this kind of activity."

FROST: But in the actual period before the Fielding break-in in August, I think, Mr. Ehrlichman has said that he talked to you about the fact that Hunt and Liddy had to go operational . . . had to go to California to find information. How much did he tell you at that time?

NIXON: I do not recall a conversation with Ehrlichman on that.

The circumstances here are interesting. Investigators found a memorandum signed by Ehrlichman approving a "covert operation" to examine Ellsberg's psychiatric files. Despite this, at his 1974 trial, Ehrlichman's defense was not that the operation had had Nixon's blessing but that he had never given *his* okay to Krogh or the other plumbers. Our suspicion was that Ehrlichman was angling for a pardon from Nixon and, until the day the president resigned, thought he was going to get one and thus had no interest in subjecting Nixon to more trouble than he was already in. It became clear to us in short order that without the prod of White House tape transcripts or credible testimony from the likes of John Dean, Nixon would have volunteered nothing.

FROST: And after the event, when he got filled in on the fact that they had done the job and found very little in this—

NIXON: Nothing—

FROST: —illegal break-in . . . nothing—

NIXON: —dry hole—

FROST: —did he, what did he tell you about it afterward?

NIXON: Nothing . . . that I can recall. I have no recollection whatever of having been told anything about it afterwards . . .

The Nixon tapes released to date show no direct connection between Nixon and the Fielding break-in. Earlier tapes did indicate Nixon's deep interest in the contents of a safe at the Brookings Institution, a think tank located in Washington, D.C., and a suggestion, attributed to Colson, that the building be firebombed and the papers retrieved. At the time of our interview no tapes had as yet captured Nixon issuing a direct order regarding Brookings, and our questions about that order elicited a cautious response.

"I have no recollection of authorizing a break-in at Brookings. I, however, would not say that I did not express deep concern about the fact that Brookings might have this and that I did not express a very great interest in trying to obtain those documents from Brookings. Ah . . . get them back in some way if we possibly could."

As always, right below the surface was resentment for having been singled out for conduct he claimed was widely practiced. Why had the plumbers run afoul of the law so quickly? But he did remember that Martin Luther King, Jr., had been bugged by the Kennedys, LBJ had had the FBI bug the Mississippi Freedom Democratic Party at the 1964 national convention, and Bobby Kennedy, as attorney general, had had reporters awakened at midnight to answer questions about monopolistic collusion in the steel industry.

If I had to pick our most enjoyable segment, it was the one in which we discussed the subject of Spiro Agnew's resignation under fire in the midst of Nixon's own Watergate travails. All of us expected Nixon to show great irritation with Agnew, who falsely protested his innocence and assured Nixon he would survive the charges. In point of fact, Nixon seemed almost relaxed about his vice president's foibles, suggesting that Attorney General Elliot Richardson—who later resigned over the Saturday Night Massacre—may have carried a personal grudge against Agnew. As for Agnew's graft as governor of Maryland, the president seemed positively tolerant. After all, demanding money from contractors who wished to be considered for state construction contracts "was common practice in most of the eastern states and many of the southern states. It had not yet swept out to the West." Furthermore, "there was never an instance when a contract did not go to a highly qualified, and, in his view, the most qualified individual. In other words, his point being that he did not, in effect, accept money from somebody who would not have otherwise been entitled to a contract." Moments later, Nixon again tried to define Agnew's philosophy of government. "I do not think for one minute that Spiro Agnew, for example, consciously felt that he was violating the law and basically that he was being bribed to do something which was wrong . . . because of a payment."

Had he thought Agnew was guilty? "I was very pragmatic," he replied. "In my view, it didn't really make any difference." There wasn't any question that Agnew "was going to get it."

Like Nixon himself, Agnew had been the victim of a "double standard." "I would say that because he was conservative, because he was one who took on the press, he got a lot rougher treatment than would have been the case had he been one of the liberals' favorite pinup boys."

In the production trailer, Birt and Zelnick alternated between bouts of hilarity and incredulity at Nixon's response. It was not Nixon's treatment of the Agnew matter at the time that they found hard to comprehend. But how could Nixon now justify defending a man who had misled him so blatantly when confronted with the charges against him? And why would he—trying to regain a semblance of national respectability—line up behind a man who was literally without a constituent in the country?

Nixon's response later was a simple one: "I just didn't want to kick him when he's down."

To many, Agnew and the Ellsberg break-in have in common the fact that both were intended for use against the liberal media establishment. Throughout his presidency, Nixon was accused of intimidating the media, his tactics having a "chilling" effect on free speech and expression. No charge, not even participation in the Watergate crimes, elicited a more heartfelt response from him. "Now, when we talk about the great period of repression during the so-called Nixon years, who was repressed? My God, was CBS repressed? Was ABC repressed? *The New York Times*? *The Washington Post*? What about the dissenters? Were they repressed? Were they afraid to speak? What was the situation? All it did as far as they were concerned was to build up their lecture fees and so forth and so on, those that claim they were depressed—repressed."

Gerald Ford was not Nixon's first choice to succeed Spiro Agnew. But the president was told that John Connally, his first choice, would have difficulty winning confirmation, as would Henry Kissinger's first choice, Nelson Rockefeller. So Ford breezed through the process and, when he succeeded Nixon, assured the nation that "our long national nightmare is over." But he soon found that our even longer national obsession with Richard Nixon was not. Knowing that an indictment and trial would keep Nixon on the front pages for the remainder of his own term and then some, Ford issued a pardon. His lawyers had researched the subject and concluded that acceptance of a pardon carries with it an implicit acknowledgement of guilt. Nixon, of course, had his own notions of the law and would eventually go to his grave without linking acceptance of the pardon to any confession of guilt. Ford would lose the presidency in 1976, most analysts concluding that the Nixon pardon had cost him the White House. More than a quarter century later, his act would win him a Profile in Courage Award from the John F. Kennedy Library Foundation.

For Nixon, the pardon, his narrow but decisive victory over phlebitis, and the Frost/Nixon interviews were the essential preliminary steps toward his return to society as an elder statesman and foreign policy expert.

I asked Nixon whether in any sense he felt that resignation had been worse than death.

"In some ways," he replied. Not as the "popular mythologists" of the period might describe it. He had had no desire to "fall on a sword" or to "take a gun and shoot myself." No. "I never think in suicidal terms, death wish and all that . . . that's all just . . . just bunk."

Resignation, however, had meant a "life without purpose," a life without the ability to contribute to the causes he believed in, to fight the battles he enjoyed fighting. In that sense, resignation had been "almost unbearable, a very shattering experience, which it has been and, to a certain extent, still is."

Again, the popular misconception of his fate. People might envy his ability to live in a nice house, wear decent clothes, play golf whenever he pleased. "And the answer is, if you don't have those things, then they can mean a great deal. When you do have them, they mean nothing to you . . .

"To me, the unhappiest people of the world are those in the watering places, the international watering places," he continued. "Like the south coast of France and Newport and Palm Springs and Palm Beach . . . going to parties every night, playing golf every afternoon, then bridge . . . drinking too much, talking too much, thinking too little. Retired. No purpose."

The people who envy that form of existence, "They don't know life. Because what makes life mean something is purpose. A goal. The battle. The struggle. Even if you don't win it." So I asked Nixon what was now his ambition, his purpose. He replied that it was to finish his book, which, along with these interviews, he hoped would "give some lessons to people in the future."

"You know, there are many famous Confucius sayings, and one of my favorites is 'One who makes a mistake and does not correct it makes another mistake.' And I suppose that is the lesson of Watergate; it's the lesson of life."

3

PRIME TIME

Bob Zelnick was relaxing in his room at the Beverly Hilton hotel less than a week before our first broadcast when the telephone rang. The caller identified herself as Nancy Collins, the author of *The Washington Post*'s gossip column. Ms. Collins quickly got to the point: Was David Frost going to confront Nixon with new tapes unearthed by his investigators? Zelnick stalled with an ambiguous reply, wondering how Ms. Collins knew about the tapes. Her next question, pertaining to the content of our planned questions, was a dead giveaway that she had somehow acquired copies of all our documents dealing with research and interview strategy completed before we left Washington for California. Zelnick's mind raced. The only possible culprit, he thought, was a secretary hired for the project who had quit when we failed to invite her to accompany us to the West Coast. He was dead-on. The woman had dumped every scrap of paper she had on *The Washington Post.* Thank God she doesn't have the transcripts of our interviews conducted at San Clemente, thought Zelnick. Regardless, there was no reason to stonewall.

The *Post*'s great national correspondent Haynes Johnson soon joined the conversation, and Zelnick tried to acquaint them with the theory of our exercise. "We looked upon the Watergate taping as the trial Nixon never had," said Zelnick. "We tried to look at ourselves as senior litigation partners in a law firm. But we knew we could ask questions and draw legal conclusions at times that a prosecutor can't draw in court. We knew we could use certain internal legal analyses and blast him with it. We were in close touch, you know, with many of the people who had prosecuted the Watergate trial." In fact, Zelnick himself had discussed certain legal theories with the prosecutors Richard Ben-Veniste and James Neal.

The *Post* played the story straight and, while we were disappointed that our control over the editorial flow had been interrupted, we could see certain public relations advantages to the piece. Other media, suspecting that the entire affair had been the product of David Frost hype, suggested that the leak had been intentional. Zelnick accepted interview requests from NBC News and National Public Radio, both of which aired stories reflecting deep skepticism about his assurance that we had not been involved. "Don't take it too hard," John Birt advised Zelnick. "Any reporter covered by the press quickly revises his opinion regarding their competence and objectivity."

In fact, the project had been buffeted by criticism since the moment it had first been conceived. There was nothing about the venture that didn't seem to rankle one columnist or another: that Nixon was being paid, that he was being given a platform, that a pushover Brit would be his only interrogator. Writing for *The Washington Post,* Charles B. Seib spoke for many of his colleagues at the time the contract was first signed: "Richard Nixon has had the last laugh in his long and bitter feud with the press.

"By selling his story of his presidency and its disastrous end to David Frost and friends he has rejected the news business in favor of show business."

For better or for worse, everyone involved with the Frost/Nixon interviews knew they would be historic. No president had ever subjected himself to anywhere near the twenty-four hours of searching inquiry represented by our project. No president had ever yielded complete editorial control to his inquisitor. And, of course, none had ever left a presidency the way Richard Nixon had or maintained silence, shunning every opportunity to communicate with the outside world, the way Richard Nixon had. At least until such time as he chose to speak again—which he didn't—or until new tapes were put into the public realm, these interviews, along with Nixon's memoir, would stand as the history of his presidency. Yes, they might in part be self-serving. But the editorial arrangements and the research conducted by our staff gave them an authority that was recognized by most media critics and that withstood the test of the next thirty years.

The events of those weeks leading up to our first broadcast still leave me dizzy. Zelnick flew back to Washington to prepare for the Wednesday afternoon embargoed screening of the Watergate program. Jim Reston headed for North Carolina. The rest of the team remained in L.A., preparing for a thank-you party at Chasen's, where we would also watch the program as it aired in L.A. on KTTV. John and I probably watched the final cut a hundred times, always with amazement at its impact and sense of drama.

Still, we did not escape some of the plagues that visit productions on opening night. In a small ballroom filled with the national press, including Woodward and Bernstein, Zelnick couldn't get his videotape to play. (He promised the group to have the problem fixed within eighteen and a half minutes.) Once it rolled, the program had an enormous impact. But that was offset when Zelnick confessed that he had accommodated AP and UPI by relaxing their embargo. A San Diego paper followed suit, and soon details about the content of our broadcast were pouring from every news facility in the country. But we were on a roll. Word was spreading that the program was outstanding and shouldn't be missed. Not even a West Coast technician's glitch, which sent the first minute or two of the program out without sound, could break the spell. We had won the greatest gamble of our lives. I thought my dad, a great but poor country preacher, must be smiling in Heaven.

Following the broadcast, Nicholas von Hoffman, the brooding sociologist turned *Washington Post* columnist, offered this assessment:

> *The Nixon-Frost interviews, far and away the best piece of journalism to pop before our eyes on the TV screen in a very long time, couldn't get on any of the networks. Nevertheless the segments have commanded large viewing audiences and been so newsworthy that the networks have been put in the embarrassing position of having to report on the very same material they had refused to put on the air. . . . Frost not only asked the right questions the right way at the right time, but, and this is rare among television magpie news performers, knew when to shut up and listen. You would think, then, CBS with its traditions would have bought the Nixon-Frost program.*

According to published reports at the time, the first program dealing with Watergate, which aired May 4, 1977, drew forty-five million viewers—the largest audience ever for a news interview (*New York Times*).

A *Newsweek*-Gallup survey said that more than 69 percent of the people who saw or heard about the interview thought Nixon was guilty of obstructing justice or other crimes. Fifty-nine percent also thought Nixon was covering up some things, although a 38 to 28 percent plurality felt more sympathetic to Nixon after the program.

Ironically, Nixon had more to gain politically by faltering under the intense pressure of an honest interview than would have been the case had he simply stonewalled. In his brilliant book *Nixon's*

Shadow: The History of an Image, David Greenberg traced the development of Nixon's public image, from his first campaign in 1946 to his death in 1994. He noted that by 1970, Nixon was largely perceived as stiff, awkward, aloof. Episodes like his staged stroll down a California beach were parodied, as in a Jules Feiffer cartoon in which a Nixon action figure is accompanied by the caption "He walks! He talks! He moves his arms up and down! He changes Expressions! The New Improved Dickie Doll! You'd almost think he's real." Watergate only exacerbated Nixon's remoteness, and by the time he left office the man was for many not a man at all but rather a distant caricature of calculated corruption. If he wanted to reenter public life, he would need to reconnect with his softer side. The interviews were an opportunity for him to demonstrate that he was indeed human.

Though Nixon's early responses were vague, evasive, defensive, by the end of our sessions I felt we had achieved more than one moment of startling honesty. This was noted by several journalists and students of the presidency. James M. Naughton, who had covered Nixon's presidency for *The New York Times,* wrote as follows:

> *But after Mr. Frost urged him to admit "wrongdoings" and apologize for it lest he be "haunted for the rest of your life" the sixty-four-year-old former president, his ruddy face wrenched by apparent emotion, and his eyes moist said:*
>
> *"I let down my friends, I let down the country. I let down our system of government and the dreams of all those young people that ought to get into government but think it's all too corrupt and the rest. I let the American people down, and I have to carry that burden with me for the rest of my life."*

Nixon, I'm sure, would have scorned the notion of being pitied. But, perhaps ironically for this man, who championed emotional reserve above all else, such brief displays of feeling would set the stage for his eventual comeback. As Greenberg recounts, "Whatever sympathy Nixon had won by 1977, it was rooted not in revived respect, but in pity. So low had Nixon's reputation sunk that he now seemed a sorry case." Nixon at last seemed unthreatening, and, for the first time in a long time, the American people began to let down their guard. "The first Frost interview," wrote his antagonist Anthony Lewis of the *New York Times,* "made this dreadful creature seem pathetic. A new Nixon had emerged after all: fallen, humanized, no longer threatening, but a pitiable figure. This was hardly the image that Nixon wished to promote, yet it was, for now, the best he was going to get."

Still, in those early days, even the Nixon team was pleased. "From our standpoint, the interviews turned out to be a plus," recalled Ray Price. "Nixon came off as human, intelligent, serious."

More important for his later "comeback," he had faced up to Watergate and other issues in a forum over which he had no control.

The week our first program aired, *Newsweek, TV Guide,* and *Time* all featured the sessions on their covers. More than 150 stations carried the show. It was the culmination of a two-year dream for this interviewer. I never did ask Nixon if he had watched the show. My guess is that he did, though I'm quite certain he would have denied it.

Today, the interviews are still in the news. They have become the subject of a West End and Broadway play authored by Peter Morgan and a Hollywood film directed by Ron Howard. Reporters who have refamiliarized themselves with the project have tended to reach the same conclusion as did those a generation ago. Writing in *The Telegraph* of August 24, 2006, Michael Sheldon had this to say:

A provincial figure in many ways, Nixon was flattered by the attention of a man whom he saw not only as cosmopolitan and sophisticated, but also as amiable and straight forward, lacking what he believed was the seething animosity of elite American journalists and pundits. He also liked the fact that Frost was a self-made man from modest beginnings.

But then, to Nixon's surprise, the polite questioner suddenly took off the gloves and confronted him with hard questions about Vietnam and Watergate. As Nixon found himself cornered, the interview became fraught with tension. Whether it was out of desperation or design, Frost managed to make the confident, composed politician crumble in front of the cameras, and the real man emerge to say he was sorry for his failings and had no one to blame but himself. His confession, which was watched by the largest audience of any television interview in history, was nothing less than stunning in its raw and tragic simplicity. . . . What Frost succeeded in unmasking was not an ogre, but a vulnerable loser who had run out of ways to hide.

The goal of any good interviewer is truth. The truth of what happened and why, of who a man is or became. Richard Nixon and Watergate remain the most complex subjects I ever tackled. He was both a man and a legend, a caricature and a character. To many he was a villain, to others a hero, to all an enigma. The excavation of Richard Nixon will no doubt continue well into this century.

Of all the subjects we covered, four stood out both on the set and in the editing room: Vietnam, Watergate, China, and the Soviet Union. They were the defining issues of the Nixon White House, and each would call on and illuminate a different Richard Nixon.

Even in the aftermath of the events themselves—in the retelling of tales already lived—I watched the man across from me transform from determined leader to wily defendant to wise statesman. The issues would divide, unite, inspire, disillusion. They would also determine Nixon's legacy and, on a far smaller scale, the legacy of our joint interview venture.

It is fair to say that as president, Richard Nixon spent more time on Vietnam than any other issue associated with his first term. It was a war he inherited, a dilemma compounded by massive public dissent, an irritant to the developing relations with the Soviet Union and the People's Republic of China, and a test—to Nixon—of America's mettle as a nation. When Nixon went on television to justify his incursion into Cambodia, his words, while grandiose, came from his heart. He was determined that America not become the "pitiful, helpless giant" described in his speech. The question raised by our debate and his responses was whether his policies had encouraged the very results he sought to avoid.

Nixon's assessment of America's Vietnam problem was described brilliantly by Jeffrey Kimball in his book *Nixon's Vietnam War:*

> *Nixon believed that American foreign policy was in crisis. But to him, in a world of challenge and change, it was a crisis of will and understanding, power and credibility, of public order and economic health. The stalemate in Vietnam was both a cause and a symptom. The antiwar opposition, the counterculture movement, and the black revolt were, he thought, rebellions against legitimate authority, which had produced social disorder and compounded the already difficult task of global leadership. Ominously, the erosion of public and elite support for intervention in Vietnam had undermined the almost two-decade-old consensus supporting the principle of military intervention itself. The nation, he thought, was on the precipice of a "new isolationism," threatening American goals and leadership abroad.*

Our interviews illuminated some of the dark corners of Nixon's and Kissinger's approach to this dilemma—secret bombings, buggings, burglaries, and more. A careful reading of Nixon's responses to our questions may produce in many minds the conclusion that his effort to buy time politically resulted in positions that made it impossible to achieve the results on the battlefield that he needed for a successful policy.

Nixon, however, did show a successful politician's understanding of the American mind. Again deferring to Mr. Kimball:

> *To most Americans the nation's vital interests did not seem to be in jeopardy, and the purposes of this ugly war were not tangible. Nixon, however, attempted to transform this intangibility into an asset. . . . Nixon redefined the meaning of such concepts as wisdom, honor and morality by portraying the struggle as one for personal and national self-respect, patriotic loyalty, the saving of POW lives, the prevention of a communist blood-bath, and the honoring of American commitments. The final punctuation of his argument was to attach these symbols not only to support for the war, but also to his own policies in waging it.*

But as the interviews underlined, in secret negotiations conducted early in his administration, Nixon was willing to withdraw all American forces from Vietnam within forty days of an internationally supervised cease-fire, while leaving the North Vietnamese and Viet Cong "in place" in the areas of South Vietnam they then occupied. At the same time, Nixon publicly announced the first of a series of withdrawals of American combat forces, a process that would continue throughout his administration as a matter of political necessity—even during North Vietnam's massive armored offensive during the spring of 1972. The Communists were demanding, more specifically, the replacement of the Saigon government headed by General Nguyen Van Thieu and the formation of a "coalition government" with the stated mission of leading the parties to national elections. Publicly, Nixon maintained that the pace of "Vietnamization," the term invented by his secretary of defense, Melvin Laird, would be determined by the level of activity on the ground, North Vietnamese conduct at the negotiating table, and the pace of South Vietnam's improvement on the battlefield. It didn't take Hanoi very long to discover, however, that the conditions were hollow and that Nixon's desire to continue withdrawals had in fact become a domestic political necessity. That situation undermined any perceived need by North Vietnam to come forward with negotiating concessions.

It also undermined the morale of the U.S. combat forces, the grunts in the field. Soldiers will fight bravely when they believe they are fighting to win. They will fight far less gallantly once they become convinced no victory is in sight. During the many years he covered military affairs, Zelnick encountered scores of Vietnam veterans who described their profound loss of morale when the meaning of such terms as "Vietnamization" and "in-place ceasefire" became clear to them. Some took refuge in alcohol, hashish and marijuana, whoring. "Gung ho" officers sometimes found themselves shot in the back—"fragged" was the term used—by troops they were attempting to lead into combat. Direct orders to move against the enemy were ignored at will. Racial incidents became a subject of severe concern to a military establishment with an officer corps that was only 2 percent black. Once these men were pulled from combat, no one would be sending them back in—regardless of anything Hanoi did to violate the accord.

Nixon told us he had hoped to engage North Vietnam's two big sponsors, the Soviet Union and the People's Republic of China, in efforts to move the negotiations off dead center. Nixon and Kissinger termed that policy "linkage." He was forced to concede, however, that neither Communist power was willing to play that game. Indeed, as Nixon's overriding goal turned to taking advantage of the Sino-Soviet split through détente with the Russians and the beginning of normalization with the Chinese, Vietnam became a dysfunctional area of U.S. policy.

Nixon recounted that the Soviets might have cooperated on a single occasion. During the 1972 summit, Brezhnev approached him and suggested that he would send Soviet President Podgorny to Hanoi to determine whether there was any flexibility in the North Vietnamese position. Shortly after that visit, North Vietnam did in fact amend its negotiating position, dropping its insistence on the removal of the Thieu government as a condition precedent to a cease-fire and the return of American POWs.

Nixon was forthcoming in describing the major explosion in South Vietnam as he and Kissinger tried to force what President Thieu regarded as a disastrous agreement on the South Vietnamese. Indeed, there can be no clearer indication of which side won and which side lost in the game of negotiations than the fact that while Kissinger was trying to plead the agreement's case with Mr. Thieu, Hanoi was holding a press conference announcing the specific terms of the agreement and demanding that Washington sign as it had pledged to do. Instead, Nixon made two relatively minor changes in the original Paris deal—both of which had been demanded by Mr. Thieu. And when the North Vietnamese expressed reluctance to let Washington go back on its word, Nixon began the infamous "Christmas bombing," a term with which he took issue, given the fact that air operations had been suspended during and just after Christmas Day.

Nixon insisted that he had achieved "peace with honor," in that the North Vietnamese had agreed to a cease-fire throughout Indochina, American POWs had come home, and our allies in Saigon were still in power, with the weapons to defend themselves. This situation would change dramatically over the next two years as Nixon's presidency became increasingly crippled by Watergate and pent-up congressional frustration at having been led into a disastrous war asserted itself. With political pressure mounting, Nixon was forced to agree to an August 15, 1973, cutoff of U.S. bombing operations throughout Indochina. He then saw congressional enactment of the War Powers Act, permitting—with certain exceptions—the president to initiate and maintain combat for a period limited to sixty days without specific congressional authorization.

The Paris Accord allowed both sides to replace their depleted forces on a one-for-one basis. But as Congress whittled down the funds for South Vietnamese military assistance, Saigon's military advantage over Hanoi increasingly became a deficit. Nixon was no longer president when, in the spring of 1975, North Vietnamese forces surged across the demilitarized zone and down the old Ho Chi Minh Trail in Laos and Cambodia, quickly routing South Vietnamese forces and completing the Communist takeover of each of the three countries of Indochina. For this, Nixon expressed bitterness toward those who had failed to enforce a deal he considered in the U.S. national interest. He maintained, as he did in a subsequent book, *No More Vietnams,* that had Congress acted responsibly, our South Vietnamese friends would have survived. He further argued that had he been president with no political shackles restraining his ability to conduct policy, Hanoi would not have had the temerity to attack South Vietnam, risking what Nixon said would have been a swift and "massive" American military response.

It was difficult for my team and me to believe that even Nixon, armed with an overwhelming four-year mandate and suffering no Watergate-related disability, could have pulled off a defense of the terms of the accord, potentially suffering additional American dead and wounded, not to mention the sight of new American POWs displayed before world cameras by the politically sharp North Vietnamese regime.

Nor, we believed, could Nixon have offered the carrots that

were also necessary to make the deal work, including $3 billion in "reconstruction" assistance to North Vietnam. Unlike World War II, after which the United States had displayed a victor's magnanimity, here it was being asked to hand over $3 billion in extortion money to an enemy that had held its own on the battlefield. Fat chance.

What was painfully clear was that Nixon had been dealt a losing hand by Lyndon Johnson and a group of military tacticians who thought they had deciphered the mysteries of counterinsurgency warfare. He played his hand well, but in the end, the other side learned what his priorities were and was able to interpret his toughest moves as bluff and bluster.

When the definitive history of the United States in Vietnam is written, Nixon will be seen as a man who realized he had more important items on his agenda but who took a shot at salvaging Vietnam and failed. His point—that Congress acted shamefully—was well documented in the Interviews. But no Congress would have permitted Nixon, or any other president, to extend this dreadful American misadventure.

Nixon's visit to China was a diplomatic extravaganza with profound policy effects, few of which—fortunately—had to be sorted out at its inception. The actual question of recognition, the approach of both powers to Taiwan, and the degree to which China's new window to the West would spark domestic political and economic reform were all matters that could be left for another day. While some conservatives groused about the reversal in U.S. attitudes, the move itself attracted overwhelming political support and contributed to Mr. Nixon's huge lead in the polls as his campaign for reelection commenced.

His policy regarding the USSR was far more complex and controversial, and, at the end of the day, he would have a good deal less to show for it. Essentially, the policy had two elements. The first was détente, a way of conducting international policy so as not to threaten each other's vital interests. The backbone of détente was a focus on keeping lines of communication open to prevent grave misunderstandings and reinforcing this approach with a network of economic, tourist, and exchange programs aimed at building support for each power's participation in the process.

The second was arms control. During the interviews Nixon vigorously defended his pursuit of détente and its harvest of agreements to limit nuclear weapons and defensive systems. Nixon was extremely articulate in describing the way arms control agreements could contribute to the kind of deterrence that would make it unlikely for either power to launch an anticipatory strike against the other. The theory was simple: Limit defensive systems so that even a massive first strike would leave the other power with sufficient weaponry to inflict unacceptable damage on the party that fired first. At the same time, place limits on offensive missile launchers with the goal of eventually reducing the number of warheads with which each power had to deal, thus making the arms race both less expensive and less likely to result in misunderstanding during a crisis.

From the outset, hard-liners despised the Nixon-Kissinger approach to the Soviets, suggesting that Nixon was being led astray by a Spenglerian pessimist whose desperate pursuit of stability would in fact produce the reverse. The Soviet Union, they argued, led by its cadre of vodka-belching thugs, was not a fit partner in efforts to defuse international tension. What Kissinger was in effect doing was conceding the West's military inferiority and seeking the best deal he could get with an irresponsible partner. Arms control followed the same pattern, the hard-liners declared. Why else would our negotiating team yield to the Soviets' numerical advantage with limits on both land- and submarine-based missile launchers? The debate would continue for years following Nixon's departure from the White House and Kissinger's from government. Ironically, by the time it ended, both Nixon and Kissinger had switched sides.

The hard-liners' case was nowhere argued more persuasively than by Paul Nitze, a man with impressive credentials as a strategic thinker and arms controller who found himself disenchanted with the Nixon-Kissinger approach. In an article published in the January 1976 issue of *Foreign Affairs,* Nitze went public with arguments he had been making to the arms control community for several years. He noted that the Soviets were spending considerable resources on antiballistic missiles as well as civil defense and launchers with the kind of throw weight needed to orbit large numbers of warheads. The Soviets' purpose, he surmised, was to develop a "nuclear war–fighting capability." This meant outdistancing the United States in both offensive and defensive systems, something that would undermine "crisis stability" because it would provide the Russians with an advantage in weaponry at each stage of escalation, leaving the United States with unacceptable alternatives: to give up or face defeat.

The critics cited the Soviets' mischief in places such as Africa, Vietnam, and the Middle East as an indication that their military initiatives were already starting to produce geopolitical results. They were joined by others with more parochial interests, such as the Jewish community—which had no problem with détente so long as it resulted in greatly increased Jewish emigration.

Support for Nixon's détente suffered a perilous blow as a result of the Yom Kippur War, when Egypt and Syria—two Soviet clients—launched a major surprise attack against Israel, nearly defeating the Jewish state in the early hours of the conflict. Before the crisis passed, Brezhnev had threatened to take action unless the Israelis suspended their effort to destroy Egypt's Third Army, and Nixon had responded with a communication of his own, plus an order to put U.S. forces on a worldwide nuclear alert. The matter came up at an October 26 presidential press conference when Nixon was questioned by the reporter David Theis:

> Q: Mr. President, against this background of détente, Mr. Brezhnev's note to you has been described as rough or perhaps brutal by one senator. Can you characterize it for us and for history in any way you can mention?
>
> A: Yes, I could characterize it, but, Mr. Theis, it wouldn't be in that national interest to do so. My notes to him he might characterize as being rather rough. However, I would rather—perhaps it would be best to characterize it. Rather than saying, Mr. Theis, that his note to me was rough and brutal, I would say that it was very firm and it left very little to the imagination as to what he intended. And my response was also very firm and left little to the imagination of how we would react. And it is because he and I know each other and it is because we have had this personal contact, that notes exchanged in that way result in a settlement rather than a confrontation.

To our surprise, Nixon made it clear in the interviews that Brezhnev had in fact warned him, while the two were visiting San Clemente during their 1973 summit, that the Egyptians and Syrians would soon launch an attack unless Israel returned to its pre-1967 lines. Nixon claimed that he had replied that in such an eventuality the United States would support its friends. But he apparently took no steps to defuse the crisis by approaching the Israelis.

Nixon was powerful enough to win this first round of confrontation with the hard-liners. But it would be the last one any president would win until the Soviet Union itself went out of business. The issues articulated by Nixon, Kissinger, and their hard-line opponents would encounter increasing opposition in the years ahead. For example, many hard-liners regarded the notion of mutual deterrence as a virtual suicide pact that no nation should voluntarily accept. Instead of an arms control regime, they urged the United States to concentrate on defensive systems to shield its population in the event of a nuclear strike. This philosophy was embraced by President Ronald Reagan in his 1983 talk introducing the Strategic Defense Initiative, or SDI, known in some circles as "Star Wars."

The notion here was to withdraw from the ABM treaty and deploy large numbers of sensors capable of detecting a Soviet missile launch in space. The system would then launch attacks intended to disable the so-called weapons bus before it released its cargo of missiles and warheads. Other interceptors would be stationed along the route, the hope being to destroy all, or nearly all, offensive weapons before they reached targets in the United States.

When adequately digested, the SDI approach was among the most bizarre ever to invade strategic doctrine.

The critical sensors would be easy to blind and easy to hit. Other interceptors could be sent on a wild-goose chase in pursuit of dummy missiles. And, in any event, the system could be overcome by firing more offensive devices than there were weapons to intercept them, so the system would at best produce an offensive missile race likely to be strategically neutral but hugely expensive.

The hard-liners also tended to exaggerate Soviet successes and Western defeats. Yes, Vietnam had ended badly, and before long the Russians would be making mischief in Africa and Afghanistan. But the Soviet Union was also a country with an economic system near collapse, a growing nationality problem, a declining life expectancy, a declining birthrate, a fearsome alcohol problem, and a deep-bred cynicism about the Communist Party and its insipid slogans. And it was a country that had suffered the most significant strategic defeat of any superpower since the 1940s when China had defected and decided to explore closer economic and political ties with the West. Not to mention that the Soviets had had to go into Czechoslovakia in 1968 in order to preserve Communist rule and that by 1980 they were demanding a crackdown by the Polish government to quash the rising Solidarity movement. Indeed, by the mid-1970s the Soviets' leading Middle Eastern client, Egypt, had sent its Russian advisers packing, recognized Israel, and tied its future to U.S. support.

Surprisingly, for all their mistakes and exaggerations, it was the hard-line opponents of Richard Nixon and Henry Kissinger who were able to claim credit for victory in the Cold War. What appears to have happened is that the Russians could not bring themselves to believe that any president would be stupid enough to embrace SDI without having the technological wherewithal to make it work. This, moreover, came at a time when Soviet military leaders were already lamenting improvements in such areas as target acquisition, accuracy, mobility, and lethality that, they claimed, made Western conventional weapons almost as efficacious as nuclear weapons. Their concern was that, by contrast, the Soviet economic system was not producing the quality of technological expertise necessary to remain competitive with the West. So an institution that had previously resisted political reform now embraced Mikhail Gorbachev's call for glasnost and perestroika, openness and restructuring. Not too many knew at the time that the system was poised to crumble, but it did. The hard-liners in the United States claimed responsibility for the result and argued that it had earned them a prominent place at future policy tables. Nixon and Kissinger eventually endorsed SDI, Kissinger on the grounds that it was "immoral" to have the ability to protect civilian populations from nuclear attack but still fail to do it; Nixon on the grounds that a modest SDI system could, along with other measures, offer protection against a limited Soviet strike.

The interviews presented a notable opportunity for Nixon to offer an early defense of policies that were ultimately rejected by his own party and those thought of as hard-liners or "neoconservatives."

I think Nixon and Kissinger each made compromises to maintain their good standing with the growing hard-line faction. But I will always remember Nixon's words in the interview when he was discussing numbers of weapons and said, "There comes a point when they don't make that much difference. Because, unless you can strike . . . knock out your potential opponent's ability to have a counterstrike, that gives you unacceptable damage. That, in other words, it kills forty, fifty, sixty, seventy million of your own people and even were a leader insane, he cannot risk losing up to a hundred million of his own people in a second strike . . . that's not an option."

With the interviews over and aired, I presumed my relationship with Richard Nixon might fade into history. But Nixon never lost his power to amaze me, even from the grave. For thirty years I have thought warmly about the rapport we established during parts of the Frost/Nixon interviews, the many delicate items we explored, and the end result, which in fact permitted him to purge some of the demons in his soul. Without the interviews or something very much like them, Nixon's final comeback—though never total—might never have taken place at all. I was satisfied with this result because it was the product of honest work and intense preparation. Our relationship during the interviews was at times symbiotic and at times adversarial. But always there was in my mind the notion that a product done with integrity would benefit everybody.

During the years following the interviews, I was told by mutual acquaintances that the former president felt that way too, that he had regarded our questions as "tough" but "fair," and that he was now ready to reenter political society.

Lo and behold, sixteen years later Mr. Nixon attacked our project in his book, *In the Arena: A Memoir of Victory, Defeat and Renewal.* Retracing his return to society, Nixon wrote about completing his memoirs and addressing Watergate issues extensively in the televised interviews conducted by yours truly. Mr. Nixon relates:

> *I agreed to make the broadcasts not by choice but by necessity. I faced a major financial crunch because of attorneys' fees. The entire amount I received from the broadcasts of $540,000 went to my lawyers. The weeks of preparing for and the twenty-six hours of taping the broadcasts proved to be the major ordeal of my stay in San Clemente. Writing my memoirs required me to engage in detached analysis and intense concentration. The Frost interviews required me to gird for intellectual combat. I did not expect the telecasts to be positive or even balanced, and I was not surprised when they turned out to be highly negative. It was a commercial enterprise, and these do not pay off by producing enlightening discussion but by producing clashes between personalities. I vividly recall my first meeting with the British media magnate, Sir James Goldsmith, who visited me while I was taping one of the programs. He was a strong supporter and was shocked by what he considered to be the vicious anti-Nixon bias of Frost's top researchers, James Reston, Jr., and Bob Zelnick, now Pentagon correspondent for ABC News. I knew he was right, and the choice of topics, the slant in the questions, and the editing for the final broadcasts reflected the bias. At the time, however, I had no choice in such matters.*

I must say that I was not apprised of Mr. Nixon's attack on myself or my staff at the time his book was published in 1990. I discovered them only while preparing this current book to commemorate the thirtieth anniversary of our interviews and the arrival of *Frost/Nixon*. During the interval, of course, Mr. Nixon died, and I have no desire to continue in death a fight that I was truly unaware of during the former president's life. Suffice it to say I was extremely proud of my staff, I thought the interviews were fairly conducted, and I gave Mr. Nixon a chance to explain Watergate, if he could. At the same time, there is something almost perversely nostalgic about seeing the "old Nixon" suddenly reappear, riding vainly into another misconceived battle.

As surprised as I was by his cracks at my staff, I was even more astonished at Mr. Nixon's reference to the late Jimmy Goldsmith, a great friend of mine and an investor in the project, who was in San Clemente at *my* invitation, when he met Richard Nixon for the first time. He had nothing but the most complimentary comments about our preparation for and conduct of the interviews. Jimmy was also a great admirer of the former president's foreign policy—they later became friends—and so was doubly pleased by the extensive discussion of international issues in the interviews themselves, plus the additional material we recorded specifically for our British audience. Let there be no doubt: the Frost/Nixon interviews were as fair in both their conduct and their editing as we poor human beings could make them. As is the case with almost all television editing, the choices we made were designed to offer viewers the most interesting material and the most important, coherently organized and logically presented. Indeed, over the years, I have been stopped and congratulated by almost as many Nixon stalwarts as opponents. To change the *Frost/Nixon* interviews, the former president would have had to change the life explored. And there's the rub.

4

THE ROAD BACK

The second half of the 1970s was not a happy time for the United States. Vietnam, the event that had defined the ethos, the culture, and the politics of the 1960s, came to a sorry end, with our allies clinging to helicopter skids or packing their families into crude and leaky boats, seeking liberty but gaining mostly death. The Russians, professing détente but playing some very familiar power games, were on the move everywhere from Angola to Mozambique to Afghanistan. And little Israel had again overcome two Soviet-backed heavies, Egypt and Syria, ensuring its continued survival.

But just barely. And the Arab world had retaliated against the United States for supporting the "Zionist entity" by announcing an embargo of oil destined for U.S. ports. Americans, who in their business and recreational habits tended to treat oil as costing zero, suddenly found themselves short of the precious commodity, suffering simultaneously sticker shock on the one hand and line rage on the other.

Even moves that bore the stamp of U.S. initiative seemed more accommodations to reality than displays of superpower prerogative. Détente, rather than a show of U.S. strength, appeared an effort to check the worldwide Soviet momentum in exchange for membership in a magic kingdom where wheat could be had at bargain prices and twelve ounces of Stolichnaya fetched an equal amount of Fanta. In arms control, too, the United States seemed inclined to accept a modest Soviet lead in both offensive and defensive systems in order to halt a race that would produce a far wider disparity.

Here and there a ray of light did shine through. The European states—particularly France and Italy—finally beat back efforts by Communist parties to win a cabinet seat or two in their respective governments, a development that would have given "Eurocommunism" a mighty push and that might also have played havoc with NATO. The Russians got something they had wanted as well: signatures by the NATO countries on an agreement known as the Helsinki Accords, which pledged all parties not to employ force to change the postwar boundaries of Europe. To many American hard-liners, this was Henry Kissinger's final act of Metternichian amorality—de facto acceptance of a totalitarian order imposed on defenseless populations by Soviet bayonets.

Never would they receive Nixon's temperamental genius back into their good graces. A former Kissinger NSC colleague, Hal Sonnenfeldt, wondered what all the fuss was about. After all, with the passage of thirty years since the war had ended, the relationship between the Soviet Union and its sister socialist states had become "organic."

Domestically, the country was also at loose ends. The racial issue, so elemental when the question was legal segregation versus equality under the law and the goal was the sort of unity of citizenship articulated in Martin Luther King, Jr.'s dream speech, had become something far more nuanced when quotas and set-asides replaced equal opportunity, "We want Black Power" replaced "We shall overcome," and Bobby Seale and his Black Panthers struck terror into the hearts of whites no less palpably than the white-hooded rabble that called itself the Ku Klux Klan had terrorized blacks in the deep South. Reflecting the views of the man who had named them to the Supreme Court, the Nixon justices had voted to end de jure segregation in the southern schools. But they drew the line against forced busing across multiple districts to achieve racial balance in areas of de facto segregation.

Meanwhile, for much of the decade, inflation and unemployment were both on the rise, introducing such terms as "stagflation" and "misery index" into the political-economic lexicon. Having enjoyed the prosperity of the World War II victor, whose industrial infrastructure had never faced attack, the United States now faced new economic challenges from Europe and Japan. During Nixon's presidency, the United States did not always find itself in a competitive position. Toyota produced cars that outperformed Fords and Chevys; Mercedes-Benz more than matched Cadillac's appeal for the upscale driver. Increasingly, Americans watched their favorite shows on Sony television sets and shaved with electric razors made by Braun. Rather than letting market forces hold sway, Nixon, on August 15, 1971, declared a ninety-day wage and price freeze in order to gain control of an inflation rate of only 4 percent. Nearly a thousand days and four "phases," one Mideast war, and one oil embargo later, inflation was prancing along in the double digits and Nixon declared the "freeze" over. He would later write that the freeze "went against my every instinct about what is good for the American economy." He had acted because congressional Democrats had passed authorizing legislation and were screaming for him to use it.

From his hideaway in San Clemente, however, Nixon paid little attention to foreign or domestic issues during the first phase of his retirement. For the first time in his adult life, he was out of the arena, and in those early days, friends who saw Nixon described him as broke and depressed, his life without apparent purpose. It was on the West Coast, just a month and a week after leaving office, that the former president was struck by phlebitis, and a month and a week after that when a dime-sized embolism was discovered in his chest. Following surgery, he suffered a precipitate drop in blood pressure and nearly died.

"How he survived the humiliation of leaving the presidency is inconceivable," marveled Brent Scowcroft years later. "But he did. And he came back. Talk about an unbelievable will."

Scowcroft had gotten to know Nixon well, having first come to the White House as the chief military assistant. He regarded the breakthrough with China as a history-changing event and admired Nixon's comeback, even at this early stage, for its demonstration of sheer doggedness. He also understood the dark side of the man, the demons, the rages, the afternoon drinking. He remembered a fight between Nixon and Kissinger, after which a boozy Nixon told Scowcroft to put the Joint Chiefs of Staff on alert, they were going to bomb the Ho Chi Minh Trail. This was long after "peace with honor" had been declared, and Scowcroft ignored Nixon, defying a direct presidential order. The incident was never mentioned between them again.

Former members of the Nixon White House team recount that the best among them, like Scowcroft, knew when to disregard his histrionics. The tragedy of Watergate may be that not everybody did.

Never one to stay down, however, Nixon took his illness as one more enemy he would have to vanquish and was back at work as soon as his doctors would allow, zipping to and from his office in a golf cart. "He was determined to get on," said Ray Price, "to focus on his memoirs." When I asked Nixon in our interviews how he thought history would remember him, he said, "That depends who writes the history." Now Nixon set about the task, producing thousands of pages—so much material, in fact, that his publishing company would later be forced to fly in two additional editors.

In the meantime, Nixon had his "gang of four"—Gannon, Sawyer, Khachigian, and Price. A strategic thinker by nature, Nixon now found himself heading his own campaign. Only this time, the seat desired was as a member in good standing of the former presidents' club. All his efforts were geared toward this goal. And if he needed a reminder of what was at stake, there was a constant string of humiliations from the White House—funds withheld, briefings curtailed, privileges denied.

Indeed, there was nothing accidental about Nixon's effort to return to public life. On the contrary, one could argue it began the day after his resignation, when Marine Master Sergeant William Gulley arrived on the scene with eleven cartons of Nixon's personal items, surreptitiously removed from the White House. Before he was discovered, Sergeant Gulley had delivered some four hundred thousand pounds of Nixon's belongings. Litigation regarding possession of and control of access to the former president's "private" papers would span more than a decade. Nixon was ready for the fight. He might have resigned, but he was not about to roll over and play dead.

In the end, Nixon's period of physical recovery lasted more than a year and included a book contract, our interview agreement, lots of golf with Brennan, a number of talks with Kissinger, and word from Beijing via his daughter Julie Nixon Eisenhower and her husband, David, that Chairman Mao, known to be suffering from cancer, would welcome a visit by Nixon. It was just the kind of opening Nixon had hoped for. The White House was aghast. At State, Kissinger was apoplectic. Ford was having trouble enough establishing his gravitas on the world stage. The last thing anyone in the administration wanted was for Ford to be upstaged by Nixon—even if the former president could contain his penchant for making his own foreign policy.

"Of course he thought about his rehabilitation," said a close friend and adviser, Dimitri Simes, in a recent interview. But that was not the only reason he traveled. "He also felt he was doing something important and necessary to him."

Nixon arrived in Beijing on February 21, 1976. The Chinese provided transportation and a warm welcome. Nixon spent an hour and forty minutes with Mao. There was no masking Nixon's satisfaction with this, his first postpresidency trip abroad. China was a very important country, and its leaders treated him almost as a visiting chief of state. Lesser countries would soon compete for his attention. All seemed to share a sense that fell somewhere between indulgence and outrage over a statesman of Nixon's stature being ridden out of office for the most nominal of offenses. Watergate—which to so many Americans symbolized profound constitutional abuse—to many people around the world epitomized the immaturity of the United States as an international muscleman. Though the Ford administration opposed Nixon's visit, Secretary Kissinger called to request a briefing upon his return, suggesting that Lieutenant General Vernon Walters fly to the West Coast to receive it. Taking his fight for ex-president privileges to the streets, Nixon insisted on Brent Scowcroft, Ford's national security adviser. His request was granted, and for the first time in a long time, the score was 1–0, Richard Nixon.

When we discussed the China visit fairly early in our sessions, Nixon recalled that Mao was "shriveled and old. . . . But if you watched his hands—the thing I remembered . . . his hands never got old. They were very fine, delicate hands." I was intrigued by the observation, and I suddenly had a thought as to how I might satisfy Jim Reston's yearning for psychological material while still holding our story line.

"Did you feel at that moment in almost another world in the sense that you were reliving that triumphant visit of '72?"

"You know, I know those, those movies, you know, with time machines and so forth are very interesting," replied Nixon, condescendingly. "My daughter Tricia rather likes them, and Julie to a less extent. But I don't watch them and I don't read much about them." Lesson: Forget once and for all trying to get Nixon to psychoanalyze himself. The moment he feels you're putting him on the couch, he heads for the hills. There were few virtues Nixon admired more than control. He fancied himself in control of his own self and remarked that Tricia had been "just as controlled" in the final days as she had on her wedding day. As for Pat, "She's very controlled, very strong." Even the speechwriter Ray Price earned kudos from the boss for "being a controlled person who seldom shows his emotion." As for Nixon himself: "I don't like to show my emotions either." Take that, you psychohistorians!

By 1978, when Nixon published his memoirs, the country was experiencing "Oil Shock II," occasioned by the outbreak of civil war in Iran under Shah Reza Pahlavi, the United States' foremost strategic ally in the Persian Gulf region. The shah would leave the country the following January. Stricken with leukemia, he was provided sanctuary by the government of Mexico, where Nixon would visit him in Cuernavaca later that year.

Having myself interviewed the shah, in his only interview during his exile, I have often thought of how much I would have enjoyed eavesdropping on their conversation. Despite differences of religion, nationality, and political systems, I fancy the two had much in common. Each had a deep love for his country and a guiding strategic vision for its respective future. Nixon wanted to engineer a second American century. He thought the country must act to support its friends and punish its enemies, often moving through surrogates, ever wary of making commitments it was not prepared to support. The shah wanted his country to be the dominant regional power, the ally of choice of the United States, which had through a coup engineered his return to power in the early 1950s. His mantra was modernization. He sought to use his country's burgeoning oil wealth to educate its citizens, modernize its institutions and economic habits, and defeat the reactionary influence of the fundamentalist clergy. It was all in vain.

Both men made critical enemies along the way. Nixon, from his first campaigns against Jerry Voorhis and Helen Gahagan Douglas, his successful investigation of Alger Hiss, the Checkers speech, his "last press conference," his Southern Strategy, and his Watergate excesses, spent nearly thirty years touching the raw nerves of partisan Democrats, "wine-and-cheese" liberals, eastern media types, and "good government" progressives. The shah faced foes from the mosque and the bazaar, the centers of piety and tradition, plus increasingly from the victims, or families of victims, of his authoritarian excesses. Each of the two had a dark side that contributed to his political demise. Nixon believed he was above the law, while the shah believed he *was* the law. Nixon hated his enemies and set no limits in countering them. The shah saw his enemies as traitors because he saw himself as the state. Both men might have ended their crises early by moving more decisively—Nixon by burning the tapes, the shah by sweeping the streets clean of demonstrators and using all his diplomatic clout to silence the Ayatollah Khomeini's incendiary broadcasts from Paris. But in the end, both men brought themselves down, Nixon by resignation, the shah by leaving Iran for medical treatment as the revolutionaries seized power, after which he realized that there was no battle left to win. Nixon did not want to put the nation through the trauma of a Senate trial; the shah believed in his heart that, as he told me, "a ruler should not stay in power by spilling the blood of his people."

The shah was an inconvenient guest, as there was always the danger his enemies would vent their wrath on the host country. President Jimmy Carter, with U.S. hostages to worry about, offered sanctuary in the United States only on the condition he abdicate. Nixon was outraged. Under Iranian tradition, the line of succession to the Peacock Throne is relinquished only when a shah abdicates or loses his head. The shah declined Mr. Carter's hospitality and moved to Cairo, where he died in July 1980. Carter sent no official party to the funeral but did provide transportation and Secret Service support for Nixon and his son-in-law Edward Cox. Nixon delivered a short eulogy, calling the shah "a real man." His line of succession remains intact.

As time passed, Nixon's travels proliferated. He visited dozens of countries and spoke with kings, prime ministers, party chairmen, legislators, and strongmen. At eighty-one, he visited the Soviet Union with his friend Robert Ellsworth, the political director of Nixon's 1968 campaign and former ambassador to NATO. They were on a bumpy Soviet air force flight, and Nixon was weary.

"Why are you going to all this trouble?" Ellsworth asked.

"So that I can influence world affairs," replied Nixon.

As his frequent-flier miles added up, it occurred to more than one administration that Nixon was conducting his own foreign policy. Scowcroft, who remained in touch with Nixon throughout the Ford presidency and beyond, certainly heard that complaint. So did others. What he was doing was unprecedented. No one had ever said that former presidents had to recuse themselves to monasteries—normally they stay connected through places such as the Council on Foreign Relations or the annual Conference on Western Security in Munich, Germany. Nixon pursued his own agenda. He traveled so widely, maintained so many sources, and acquired so much firsthand experience that he became in effect the world's premier foreign affairs journalist.

But while his status abroad continued to flourish, little had changed for Nixon at home. By mid-1978, Nixon's situation, like the country's, seemed unsettled if not bleak. As Jonathan Aitken wrote in the biography *Nixon: A Life:*

Nixon's future in 1978, at the age of sixty-five, did not look particularly bright. He was still continuously referred to in the press as "the disgraced ex-President." The Carter White House was treating him with a pronounced meanness of spirit, restricting his briefings and other courtesies normally extended to former presidents, to the minimum. There were no indications in the polls that his low standing with the public was improving. Yet, on the wider horizons, Nixon knew he had both a domestic and an international constituency whose admiration for his achievements had not been unduly diminished by the shenanigans of Watergate. It was to these groups that he now turned in his quest for political acceptance.

All of this was about to change. Nixon had been accumulating speaking invitations—by midyear 1978, a breathtaking one hundred thousand of them. He had declined them all for fear of embarrassing his hosts, but now he was gaining confidence. He was ready to test the waters.

Nixon began in the friendly confines of Hyden, Kentucky, described by *Time* magazine as "a remote eastern Kentucky coal mining town of 500, Republican since the Civil War, where the virtue of loyalty has been toughened into a kind of clannish defiance. 'All Nixon did was stand by his friends,' said the local motel owner. 'And that is one of the traits of us mountain people.' "

Hyden was also the recipient of a Nixon-generated revenue-sharing program, which had helped the town build a $2.5 million recreation center. Landing in ninety-degree heat, Nixon found a thousand townspeople waiting for him, some sporting campaign buttons and even the thick volume housing his memoirs. As four high school musicians played "Hail to the Chief," Nixon plunged into the small crowd as though it were a pack of rural Californians replacing Congressman Jerry Voorhis with a returning naval officer named Richard Nixon. A few months later, Nixon repeated the exercise on a Veterans Day celebration in Biloxi, Mississippi.

Nixon also made his first Washington appearance that year, coming back to the capital to attend the funeral of Hubert Humphrey. Days before he died, Humphrey had phoned Nixon, and the two men had spoken at length. Ken Khachigian, still working with Nixon, described the conversation as "deep and sentimental, like two old capos recalling past battles without rancor." The trip made Nixon nervous, and his discomfort was not mitigated when President Jimmy Carter—in yet another snub—refused to invite him to stay in a Lafayette Square presidential guest facility. Nixon holed up at a friend's house in Virginia, where he considered canceling all plans but the funeral. In office, Nixon had often worried over his choices. "He would make a big decision, then go up to Camp David and agonize over whether he did the right thing," remembered Scowcroft. "He would absolutely torture himself." Now Nixon wondered if he should have even come to Washington. Following the funeral, he tried to beg off from a reception hosted by Senator Howard Baker of Tennessee, but Baker insisted he attend. Nixon relented and managed to get through the event. And on January 29, 1979, when Carter held a state dinner in Washington for visiting Chinese strongman Deng Xiaoping, he made certain that Nixon was at the table. Nixon's perseverance was paying off.

Nixon also accepted an invitation to address the Oxford Union in Britain. Despite a handful of protesters gathered outside, the audience greeted his talk on foreign policy and world affairs with a standing ovation. While in England, he met privately with Margaret Thatcher and spoke before a packed meeting room at the House of Commons. He was on a roll now, gaining confidence with each event, suggesting a reverence for Theodore Roosevelt when he told a friendly columnist, Nick Thimmesch,

> *A man is not finished when he is defeated. He is finished when he quits. My philosophy is that no matter how many times you are*

knocked down, you get off that floor even if you are bloody, battered, and beaten, and just keep slugging—providing you have something to live for. If you have something you believe in, something worth fighting for, the greatest test is not when you are standing but when you are down on that floor. You've got to get up and start banging again. Get up—and start banging again.

As if to prove he was ready for a tougher venue, Nixon began searching for a home in New York City and moved there early in 1980. He was back. Critics be damned.

Public success, for Nixon, seemed to go hand in hand with the quieting of private demons. Prior to his move, he reconciled with Bob Haldeman, freshly released from jail, and entertained 250 guests at a party honoring John Mitchell. Nixon's toast: "John Mitchell has friends, and he stands by them." His memoirs, meanwhile, though treated roughly by most critics, became an international best seller. Having saved his Watergate preparation until after our interviews, it came as no surprise to me that little of his narrative differed from his testimony at our "trial."

Nixon's New York routine has been amply chronicled: awake at 5 A.M., walking briskly along Third and Second Avenues, usually clad in a dark blue suit. His exercise completed, Nixon would get to work reading, writing, and telephoning. The day was punctuated by lunch at a fashionable restaurant, followed by dinner at home and hours of reading or watching sports and public affairs on television. "He was always strategizing—in politics, in football, in foreign policy," said Ray Price. "He was always trying to strategize the game while he was watching the game." Nixon wanted John Connally to win the 1980 GOP nomination, but George Bush emerged as Ronald Reagan's most formidable challenger, while Connally spent more money per delegate vote than any candidate in the history of presidential politics. Nixon was not yet respectable enough to attend the GOP convention or mingle publicly with party leaders, but Reagan privately sought his advice throughout the campaign, as he would again in 1984.

Nixon's second postresignation book, *The Real War,* was published in June 1980 and, like his memoirs, received bad reviews while vaulting to the top of best-seller lists. Rereading the book, even today, it is difficult to interpret it in any fashion other than as an application for membership in the same club of hard-liners who had opposed his own efforts at détente and arms control. Nixon argues that the Soviet Union, as an orthodox Communist power, is trying to take over the world and is positioning itself to do so by rapidly surpassing the West in military might. Whereas Nixon the president saw détente as a way of identifying areas of common interests and establishing a network of personal relationships that could prevent a crisis from boiling over into war, the former president now saw détente more as a trap for the gullible. Whereas the president saw additional numbers of intercontinental ballistic missiles, or ICBMs, as meaningless once deterrence had been achieved, the former president now favored the same kind of nuclear war–fighting capabilities and strategic defenses that critics used to undermine his negotiation efforts. One example: the first SALT agreement had provided the Soviets with small numerical advantages in terms of ICBMs and submarine launchers but made no mention of restrictions on intercontinental bombers, a weapon that had been in the U.S. arsenal since the dawn of the Cold War but that the Soviets had never developed. The Jackson Amendment to the accord, however, provided for equal number of launchers from all systems, including aircraft. Therefore, to reach Jacksonian equality, the United States might well have had to dismantle weapons it had earlier deployed. Stunningly, Nixon now endorsed the Jackson Amendment.

The most charitable explanation for Nixon's reversal is that since his departure, the presidency had fallen into the hands of

Jimmy Carter, an inexperienced player on the world stage whose high intelligence never translated itself into hardheaded strategic thinking. Nixon's rant against softheaded diplomacy was perhaps meant to serve as a splash of cold water in the face of this new president. On the other hand, Nixon was one of the more sophisticated political analysts of his time. Quite likely, he saw the Republican Party evolving toward a more conservative center of gravity and felt this attitude reflected in the hard-line positions toward Moscow that most Republicans were supporting.

Back on the East Coast, Nixon spent most of the 1980s generating a library of books on leadership and foreign affairs. Following his memoirs, Nixon published *The Real War; Leaders; Real Peace; No More Vietnams; 1999: Victory Without War;* and *In the Arena: A Memoir of Victory, Defeat, and Renewal.* The last was an interesting effort by the former president to set the record straight as regards the major clashes of his life and to draw from those events life lessons worthy of retention. It also stands out as Nixon's most blatant attempt to rewrite his own history.

Early in his book, Mr. Nixon recites a number of so-called myths that have come to burden his account of the Watergate episode and his involvement—or noninvolvement—in it. A fresh examination suggests that in certain of the cases, Mr. Nixon established straw men whose existence seems to be only for purposes of their destruction. Others quibble with words; still others stretch the facts. So that the record will again be clear, I should like to review Mr. Nixon's claims not only in light of the evidence that was known at the time of our interviews or acknowledged during the interviews but in the context of works by other authors using material that has come to light since 1974. Let me deal with them in the order in which they appear in Mr. Nixon's book.

1. "The most blatantly false myth," Nixon says, "was that I ordered the break-in at the Democratic Headquarters."

This hardly qualifies as a "myth" because, as best as I can determine, it has few if any adherents. No evidence to date suggests that Mr. Nixon knew of the break-in before it occurred or specifically ordered it. It would have been appropriate for the former president to inform us of where this myth was accepted. As of now it appears more a fantasy than a fact. That was why we hadn't raised the question.

2. "The most politically damaging myth was that I personally ordered the payment of money to Howard Hunt and the other original defendants to keep them silent." Mr. Nixon goes on to acknowledge that "I did discuss this possibility during a meeting with John Dean and Bob Haldeman on March 21, 1973. In the tape recording of the meeting, it is clear that I considered paying the money."

"Considered paying the money," indeed. In perhaps the most famous exchange of the interviews, I confronted Mr. Nixon with sixteen examples of moments when he not only approved of the payment of hush money to the defendants but kept pulling his associates back to the subject whenever the conversation strayed. Mr. Nixon admitted he had been told that the money was necessary to "keep the defendants on the reservation" or to keep them from "blowing." In a tradition later made famous by Bill Clinton and the verb "is," Mr. Nixon insisted that one first had to appreciate what "blowing" meant before condemning him for endorsing the plan.

Here we must restate a matter of law that seemed as alien to Mr. Nixon's thinking as it is central to the thinking of any attorney involved in a criminal case. Very simply, if you conspire to subvert justice by paying money to keep a defendant quiet or by fraudulently seeking the cooperation of one government agency to call off another, thereby limiting the scope of the investigation, you are guilty of obstructing justice—even if your motive is *only* to avert political embarrassment.

3. "The most serious myth—and the one that ultimately forced me to resign—was that on my specific orders the CIA obstructed the FBI from pursuing its criminal investigation of the Watergate break-in."

The response to this is best left to the Nixon people themselves. In his account of his association with Nixon during these trying days, Ray Price describes how he was busily working on a Nixon speech, pledging to continue his battle against impeachment, when he got word to report immediately to the office of Chief of Staff Alexander Haig, who told him, "We need a resignation speech." White House Council J. Fred Buzhardt had just listened to the infamous tape of June 23, 1972, in which Nixon tells Bob Haldeman to use the CIA to blunt the FBI Watergate investigation.

During that conversation, Nixon tells Haldeman, "When you get in these people, when you . . . get these people in, say: 'Look, the problem is that this will open the whole, the whole Bay of Pigs thing, and the president just feels that,' ah, without going into the details . . . don't, don't lie to them to the extent to say there is no involvement, but just say this is sort of a comedy of errors, bizarre, without getting into it, 'the president believes that it is going to open the whole Bay of Pigs thing up again.' And, ah, because these people are plugging for, for keeps, and that they should call the FBI in and say that we wish, for the country, don't go any further into this case, period!"

Buzhardt described the tapes as "a disaster," saying that they were the "smoking pistol" that tied Nixon directly, and at the very outset, to the cover-up. Haig and the imported White House counsel, James D. St. Clair, had read the transcripts. "They were every bit as bad as Buzhardt said they were. The tapes had to be turned over, and that meant the battle was lost." Nixon would later claim that his conduct had been purged on July 6 when, after being warned by FBI Director-Designate L. Patrick Gray that his aides were trying to "mortally wound" the president, Nixon instructed Gray to perform a thorough investigation. During the interviews, I noted for Mr. Nixon's benefit that even by his own reckoning he had participated in the cover-up from June 23 to July 6. But the more important fact is that, as mentioned above, while telling Gray to go ahead with his investigation, on the one hand, Mr. Nixon was conspiring with his top aides to keep the full story, including the involvement of Hunt and Liddy, from ever surfacing.

4. "The most preposterous myth was that I, or members of the White House staff, erased 18.5 minutes of incriminating conversation from one of the White House tapes."

No one has accused Nixon of deliberately erasing the eighteen and a half minutes, but the circumstantial evidence of wrongdoing is strong. For one thing, a panel of technical experts that examined the tape, discounted the possibility—suggested by the president's secretary, Rose Mary Woods—that the erasure was probably due to a "transcribing error." Instead, the panel found between five and nine manual erasures responsible for creating the gap. Only three people had access to the machine: Ms. Woods, President Nixon, and an aide named Steve Bull. To date, no wrongdoing has been proven with respect to the erasures. At one point Al Haig suggested that only a "devil theory" could account for the gap.

Regarding the material itself, Haldeman's notes refer only to the specific action the president ordered him to take in response to the break-in—a public relations offensive plus examining the

Executive Office Building for recording devices possibly planted by the Democrats. In his book, *The Ends of Power,* Haldeman fleshes out his conversation with Nixon in a manner distinctly less benign:

> NIXON: Colson can talk about the president, if he cracks. You know I was on Colson's tail for months to nail Larry O'Brien on the Hughes deal. Colson told me he was going to get the information I wanted one way or another. And that was O'Brien's office they were bugging, wasn't it? Colson's boy, Hunt, Christ!
>
> HALDEMAN: Still, Magruder didn't even mention Colson.
>
> NIXON: He will.
>
> HALDEMAN: Why?
>
> NIXON: Colson called him and got the whole operation started. Right from the goddamn White House. With Hunt and Liddy sitting in his lap. I just hope the FBI doesn't check the office log and put it together with that Hunt and Liddy meeting in Colson's office.

5. "The most one-sided myth was that I used government agencies *illegally* by asking Secretary of the Treasury George Shultz to order Internal Revenue Service audits of a political adversary Larry O'Brien."

Nixon admits this charge but adds, "I have no regrets for that action." He cites past instances of alleged IRS harassment of his own supporters, including the evangelist Billy Graham. Dozens of people were targeted by the Nixon administration for especially aggressive treatment by the IRS. Rather than a criminal action, a number of observers have concluded that Nixon's action was more an abuse of power and could well have formed one basis for an article of impeachment.

6. "The most hypocritical myth was that the Nixon administration sold ambassadorships to major political contributors."

Nixon appears to have carried the practice of rewarding big contributors with ambassadorships to excess. In his book *Watergate,* Fred Emery notes that thirteen ambassadors named after 1972 had contributed a total of $706,000 to the Nixon campaign. On the day in 1974 when Nixon publicly denied selling ambassadorships, Emery notes that the president's private lawyer, Herbert Kalmbach, pleaded guilty to selling ambassadorships in exchange for immunity on all other charges.

7. "The most personally disturbing myth was that I deliberately lied throughout the Watergate period in my press conferences and in my speeches."

As noted in the interviews, Nixon told the American people on August 15, 1973, that he had asked for John Dean to write a report regarding White House involvement in the Watergate cover-up. "If anybody at the White House or high up in my campaign had been involved in wrongdoing of any kind, I wanted the White House to take the lead in making that known. On March twenty-first I instructed Dean to write a complete report on all that he knew on the Watergate matter." As I noted in responding to Nixon, the president had already asked for a "self-serving goddamn statement" denying the culpability of principal figures. When Dean told Nixon that the original Liddy plan had involved bugging, "you told him to omit that fact in his document and state it was for . . . the plan was for 'totally legal' intelligence operations." The president had also told acting Attorney General Henry Petersen that Dean had produced a report that was "accurate, but it was not full." No such report had been prepared. In his public statements, the president conveyed one untruth after another, as he later admitted during our interview on April 15, 1977:

"Now under all these circumstances, my reactions in some of the statements and press conferences and so forth after that, I want to say right here and now, I said things that were not true. Most of them were fundamentally true on the big issues, but without going so far as I should have gone, and saying, perhaps, that I had considered other things but not done them."

8. "The most widely believed myth was that I ordered massive illegal wiretapping and surveillance of political opponents, members of the House and Senate, and news media reporters."

Based upon existing law, Nixon appears to be correct in that the wiretapping he ordered had legitimate national security roots. What appears to have run afoul of the law was the Huston plan, which authorized so-called black-bag jobs against Weathermen, Black Panthers, and others suspected of plotting violence. The plan's life was short if not sweet; Nixon countermanded his own order after FBI Director J. Edgar Hoover refused to cooperate.

9. "A related accusation was that I ordered members of the White House staff to arrange the break-in to the office of Daniel Ellsberg's psychiatrist, Dr. Lewis Fielding, in September 1971."

None of the Nixon tapes released to date had revealed any direct connection between Mr. Nixon and the Fielding break-in. However, tapes released in 1996 did specifically link Mr. Nixon to the order for a covert action against the Brookings Institution, but there is no evidence it was ever carried out.

10. "The most ridiculous myth was that I was the first President to tape some of my conversations."

I have not heard anyone say Nixon was the only president to tape some conversations. But evidence suggests that Nixon's system was by far the most elaborate.

11. "The most unfair myth—and the one that most angered me—was that I profited from my service as President."

Another straw man; no reputable individual has made that charge.

12. "The most vicious myth was that I tried to cheat on my income taxes."

In March 1974, the Congressional Joint Committee on Internal Revenue Taxation issued a report after Nixon asked it to review two questionable tax entries he had made on his returns for 1969 and 1970. The committee concluded that Nixon owed more than $300,000 in back taxes, resulting from his having taken a $482,000 deduction for the gift of his vice presidential papers and his failure to report a $142,000 profit on the sale of his Manhattan apartment in 1969.

Mr. Nixon conceded that the vice-presidential papers transaction might not have been completed before a law banning such deductions took effect; indeed, a deed and other papers regarding the transaction had been backdated in order to meet the deadline. Although Mr. Nixon escaped the possibility of prosecution because of President Ford's pardon, three men, including one administration official, were criminally prosecuted for their roles in the fraud.

It is hard to imagine what motivated Mr. Nixon to reopen old wounds that, for the most part, had been self-inflicted. Whether the subject was the break-in, the cover-up, payment of hush money, acts of perjury, or broader deceptions practiced on the American people, close associates of Richard Nixon went to jail for many of the things that involved the president himself. Some zest for meaningless combat, a desire to settle scores that had long since been settled, and an almost inexplicable desire to respond aggressively to perceived acts of persecution seem to have been characteristics that Nixon carried to the grave.

Other Nixon books followed. They, too, would have their critics, but once again, Richard Nixon would cry all the way to the bank, the airport, and the television studio. By the mid-1980s, he was persona grata just about everywhere. The Reagan presidential campaign, which four years earlier had secretly sought his advice, now threw the sashes back and let the starlight in. One Nixon memo surviving from that campaign advises, "The president should again make the point that during his watch not one inch of territory has come under Communist domination or been lost to the West. This is under stark contrast to what happened during the previous administration, when Ethiopia and Nicaragua came under Communist domination and Iran was lost to the West." Nixon also remained coolheaded after Reagan's slothful performance against Walter Mondale in the first of their two presidential debates. Though he felt it would have been wise for the campaign to come up with a line or two Reagan could use to defuse the question of age, Nixon urged the Reagan forces not to worry, the debates were not that important. "What this adds up to," he said, "is that debates can affect the result, but only by two or three points—not massively. With all the polls indicating that Reagan is ahead now by between 15 and 20 points, there is no way the debate by itself will tighten up the race significantly."

Clearly Nixon was reveling in his newfound celebrity. In his fine book *Nixon in Exile,* Robert Sam Anson described a typical

Nixon dinner party, strictly stag, with guest lists that included old White House aides, visiting English celebrities, and experts on the economy such as Herbert Stein, Alan Greenspan, Peter Petersen, and Arthur Burns. Conversation would be serious and profound, Nixon directing the flow with well-structured questions. The dinner was always Chinese. A Nixon dinner party, however, rarely lasted beyond 10:30 P.M., when "the host would shoot a glance at the most upright of his guests and announce, 'Well, I promised to get so-and-so to the local house of ill-repute by 11 so I guess we ought to call it a night.' "

Still, despite his celebrity status and emerging role in domestic politics, Nixon's eye was always on foreign affairs. As the decade progressed, his interest in China was overshadowed by the break-up of the Soviet Union. It was during this period that Nixon first encountered Dimitri Simes, a Soviet intellectual who had emigrated to the United States in the early 1970s. Simes was unusual in that his mother was a distinguished defense lawyer in Moscow and Simes himself was a member of the Institute of World Economic and International Relations, a sure ticket to success, status, and privilege in the Moscow of that era. Even after he emigrated, Simes stayed in touch with former colleagues, giving him an impressive insight into economic and political conditions inside the Soviet Union but cloaking Simes himself with an aura of suspicion in certain hard-line American circles—one he had difficulty expunging.

Simes and Nixon became acquainted in the mid-1980s after Simes wrote a *Christian Science Monitor* column calling Nixon a great patriot with a dark side that he had allowed to define his presidency. Nixon then sent Simes a couple of columns, which Simes critiqued with candor. After Nixon's chief of staff, John Taylor, organized a meeting in New York, Nixon suggested that Simes become a formal adviser. "At first we corresponded, then he phoned, then he asked me to phone him, then we met in person, and then at some point, he asked me how he should respond to certain events." The two would become fast friends.

Simes had two goals for Nixon: (1) he wanted him to go to the Soviet Union with some frequency so that he could develop a more intuitive feel for the place while also getting to know a range of Russians that went beyond Gorbachev and his senior advisers; and, more immediately, (2) he wanted to get Nixon speaking publicly in Washington, the *sine qua non* of a complete comeback and the most convincing measure of the scale of his comeback.

At the time, Simes was a senior fellow at the Carnegie Institute for International Peace. The irony was not lost on Nixon. Carnegie's first president had been Alger Hiss. Simes enlisted the support of Tom Hughes, Carnegie's president in the mid-1980s, and Nixon received an invitation to speak. "The first time Nixon spoke, he was brilliant and got a standing ovation," Simes recalls. "Nixon loved every second of it. You could see a certain delicious irony in doing it at Alger Hiss's place."

Nixon and Simes traveled to the Soviet Union together for the first time in 1988. Earlier in the year, talk of a coup had been dismissed by most intellectuals, who reasoned that the prize was not worth the risk. "It would be like conducting a coup in Lebanon," suggested Sergei M. Rogov, director of the Institute for the USA and Canadian Studies. But as time passed, the military lost confidence that Gorbachev's reforms would make a difference. They saw the country drifting toward unrestrained freedom, which they equated with the "rule of the mob," and thought a coup would be the only way to save communism, a system with which the military was more comfortable than were most other elements in Soviet society.

Simes and Nixon's analysis went further. They both concluded that the Soviet Union was in the process of disintegrating and that the opportunities for constructive U.S. action were many.

After Nixon returned from Moscow, he went to see Vice President George H. W. Bush. Bush extended the meeting to include James Baker and lunch. Simes recalled, "What a treat it was for Nixon. He felt like once again he was back in the arena."

The pair traveled again to Russia but this time also included Ukraine, Latvia, Poland, the Czech Republic, West Germany, East Germany, and Britain. Nixon became convinced that Gorbachev—for all his humanism—lacked the support of the Russian people and would never be able to govern effectively. Yeltsin, Gorbachev's principal rival for power, was the better bet. For all his personal weaknesses, he was more likely a democrat at heart and, in any event, was likely to prevail.

In August 1991, the coup leaders struck while Gorbachev was vacationing at a Black Sea resort. The coup quickly collapsed as the Russian people rose up against the coup masters, their courage forever symbolized by Boris Yeltsin mounting a tank and telling the military not to fire. By then it was obvious that although Gorbachev might return to office, he would never return to power.

Whoever succeeded Gorbachev, Nixon and Simes believed, would take over a country in peril. The political system would be lucky to escape civil war. The external empire had already dissolved, and the internal empire was quickly fading. The economy was structurally corrupt and was in the kind of shape that only three quarters of a century of communism could bring about. The situation was pregnant with the danger of economic collapse, institutional collapse, civil war, mass rioting, and a special peril resulting from the enormous stockpiles of nuclear weapons and equipment during a period of disintegrating central or even local authority. Nixon felt that the United States must recognize that it alone was a world superpower and that what it did with respect to Moscow during a very brief period would have a profound influence upon the question of whether there would be a "new American century" and what its condition would be. This meant providing not only political support for Yeltsin but also massive economic assistance to the Soviet Union to enable it to surmount the crisis through which it was passing. Nixon articulated these views during an April appearance on the television program *60 Minutes;* in two books, *1999: Victory Without War* and *Seize the Moment;* and in an assortment of newspaper commentaries.

In 1989, Nixon would hire his last foreign policy research assistant, Monica Crowley, fresh out of Colgate University—again following an exchange of letters. She appears to have regarded it as her mission to become Nixon's Boswell, capturing every word he uttered whether on foreign affairs, U.S. politics, the requisites of politics, the media, or the world's future. She has published two books about Nixon: *Nixon off the Record* and *Nixon in Winter.* It would be impossible for anyone to have so high a percentage of his words preserved for posterity without at times appearing boring, redundant, petty, short-tempered, and downright supercilious toward one and all. This is certainly true of Nixon. But, particularly during the period of the disintegration of the USSR, her work presents an accurate picture of Nixon's thoughts and words. Crowley describes in *Nixon in Winter* an April 12, 1990, speech to the Boston World Affairs Council in which Nixon urged an assortment of policies to help the countries that had broken away from Soviet domination:

> *He advocated only very modest defense cuts. He urged a restructuring of the North Atlantic Treaty Organization to make it more responsive to the needs of the new era. He supported opening the economic and military sectors of the European Community to the newly free nations of Eastern Europe when they showed an irreversible commitment to democracy and free markets. And above all, he warned that the United States must be prepared to aid the democratic forces in these regions or face new confrontations.*

Following the failed Moscow coup and Gorbachev's crescendoing loss of control, Nixon increasingly felt that the administration of George Bush was being far too timid in seizing the opportunities. He also grumbled that Bush was doing little to prepare for the coming presidential elections and was mishandling the economy. In *Nixon off the Record,* Crowley offers this snapshot of Nixon's attitude:

> *Goddam it! Why the hell isn't he showing some leadership? I'll tell you something. When the shit hits the fan and his gang comes to me for advice, I'm not going to provide it unless they are willing to thank me publicly. Neither Reagan nor Bush did that after all these years of my advice, and frankly I have had it. They'll find me when they need me, but I may not be available.*

Nixon underlined his feeling when, in preparation for a conference, "America's Role in the Emerging World," he wrote an essay titled "How the West Lost the Cold War," in which he described administration policies towards Russia as "penny-ante" and "pathetically inadequate." At the conference, held at the Four Seasons Hotel in Washington, D.C., amid a glittering cast of national security celebrities, Nixon gave what was widely perceived as a zinger of a speech while Bush seemed to lack completely any sense of urgency regarding events in Russia. In fact, of course, President Bush himself was rather proud of his policy toward the disintegrating Soviet empire. He quite deliberately played it low key because he felt that any hint of triumphalism on the part of the United States might provoke the Communist old guard to attempt to halt the collapse of the old regime.

Still, Nixon hoped Bush would be reelected and had no qualms about delivering unrequested memos offering the advice of a president who had carried all but one state in his quest for reelection and who had been invited to offer strategic help to Ronald Reagan in a reelection bid that had carried all but two states. Before the primary season ended, H. Ross Perot visited Nixon's home for a freewheeling political discussion and Pat Buchanan had received the accumulated wisdom of a man he once described as "like a father to me." Nixon was not asked for advice by Bill Clinton, whose character flaws he frequently grumbled about, but that did not keep him from sending the victorious Clinton a note praising him for running one of the great campaigns in history. Within months, Nixon would find himself a welcome visitor at the Clinton White House, where his advice was sought on a variety of issues and his political instincts were no less prized.

Pat Nixon died on June 22, 1993, after a long battle with lung cancer. Nixon was deeply annoyed when Clinton picked an emissary, Vernon Jordan, to attend the funeral as his representative. Those who knew the couple best say that Nixon admired his wife's coolness under fire, her dogged allegiance to him over the years, her role in raising two daughters he adored, and her consistently good judgment, especially about people. The funeral was held at the Richard Nixon Library & Birthplace Foundation in Yorba Linda four days later. Crowley described the scene as follows:

> *The small crowd stood as Mrs. Nixon's coffin was brought in and the former first family filed in behind it. Haldeman, who was seated directly in front of me, whispered suddenly, "The president's lost it." I looked at Nixon. At the sight of the casket and the guests, he had broken down and sobbed uncontrollably, shoulders hunched forward, frame trembling, tears pouring from watery eyes. The collapse in pain and sorrow became one of the day's searing images and shocked even Nixon when he saw the film of it later.*

Nixon's loneliness following Pat's death was painfully evident to those closest to him. Crowley, of course, was in touch with him daily, often for many hours. He also leaned on Dimitri Simes for companionship. "We shared many drinks by phone," Simes recalls. "My days often started with a call from him at 8 A.M. and ended with a call at 11 P.M." Their talks often concluded with Nixon describing what he was drinking, how much he had sipped, and even the fact that he was crawling into bed and turning out the light.

On March 3, 1994, Clinton called Nixon from the White House to discuss what would be the former president's final trip to Russia. Upon arrival in Moscow, Nixon met with opponents of the Yeltsin government. The man Clinton had described as the appropriate person for Washington to support retaliated for this insult by revoking Nixon's invitation to visit the Kremlin and pulling his government limousines and bodyguards. Nixon met with members of the press and expressed disappointment with Yeltsin but praised his commitment to political and economic reform. Upon returning to the United States, he dictated a memo to Clinton on the Russian situation.

One of the last pictures of Nixon shows him addressing, in animated fashion, an audience that included, among others, Les Aspin, Joseph Nye, and John Deutsch. According to Simes, Nixon held the audience spellbound for more than an hour and seemed in total control of both his material and himself. But as he walked out of the building, his legs went limp, and Simes and a colleague had to practically drag him to his car. Nixon, undoubtedly, was a man who simply would never give up.

On April 17, as he prepared to view the edits on his latest book, *Beyond Peace,* Nixon suffered a massive stroke. He died five days later in New York City.

At his funeral President Clinton said, "May the day of judging President Nixon on anything less than his entire life and career come to a close."

By any reasonable standard, Nixon's resurgence represents one of the most remarkable achievements in the history of American politics. He achieved it methodically and with patience, first by putting his financial life in order and working himself back into a position of robust health; second, by facing up to Watergate and related charges against him, as he did during many hours of interrogation during Frost/Nixon; third, by leading from strength—his relationships with Chinese and Russian leaders and insights into both societies; fourth, by moving back to the East Coast, the intellectual and political heart of the nation; and finally, by exercising self-control, playing his game within its natural limits, not trying to push the envelope further than common sense dictated.

Yet the humiliating circumstances of his departure from Washington were never far from his mind. "He was not delusional," recalled Dimitri Simes. "There are people who make great mistakes and never accept them. Nixon knew the mistakes he had made. He said to me, 'I never expect you to defend the indefensible.' "

Bob Ellsworth recalls a dinner meeting with Nixon where the former president repeated the words he used during Frost/Nixon a decade earlier, "When I asked him about Watergate, he said, 'Goddamn it, Bob, I gave my enemies a sword and they stuck it in me.' And then he shoved his napkin in his mouth and chewed on it."

With the exception of his self-aggrandizing rebuke to various Watergate "myths" in his book *In the Arena,* in the later years of his life Nixon seemed to accept the fact that his actions, or inactions, had contributed to the greatest American political scandal in history. Whether he ever appreciated the depth of the threat Watergate posed to the U.S. system of government remains a matter of some debate.

5

REEVALUATING NIXON

A dozen years after his passing, the fact that Richard Nixon remains a subject of fascination on at least four of the world's continents should surprise no one. He spent nearly half a century at or near center stage in the international arena. He was a participant in great battles involving domestic and foreign security, race and politics, the law of the jungle, and the rule of law. Five times he was nominated by his party for national office; four times he won. His greatest electoral triumph gave birth to his most ignominious defeat and disgrace as he became the first and thus far the only president forced from office by his own misconduct. Foes seemed to regard his fall as the personification of Divine Justice. Even those who remained personally loyal came to see two Nixons: one detached, analytical, a student of greatness striving to achieve peace in his time; the other, dark, coarse, vindictive, conspiratorial, a refugee from the pages of Greek tragedy, doomed to self-destruction by fatal character flaws.

Nixon can be reconsidered, even reevaluated, but he can't be reinvented. His presidency was destroyed by his own criminality, dishonesty, and ineptitude, and the country suffered through a "long national nightmare." It deserved better, much better. Still, this is a good time to consider his record as a whole. His supporters urge that the record is impressive. Pat Buchanan, his "conservative" speechwriter, calls Nixon the second-best politician of the century, claiming that he broke apart the New Deal coalition of Franklin Delano Roosevelt (Buchanan's number one), a realignment that dominated five out of six presidential elections and eventually recaptured the Congress. Ray Price, the former *New York Herald Tribune* editorial writer who became Nixon's "moderate" speechwriter, is another who places Nixon in the front rank of presidents. "Eventually, Nixon will be ranked among the better—maybe the best—of the presidents," he argues. "But not until the evaluations are made by people with no vested interest in Nixon as evil." It is Price's view that those who feverishly prosecuted Nixon for Watergate ought to have relented, thereby allowing him to conclude what would have been a second term of awesome accomplishment.

Perhaps politics is the best place to begin. During Barry Goldwater's heyday as the choice of conservative Republicans, the historian James MacGregor Burns wrote a column suggesting that Goldwater was truly radical because he sought to change the United States from a four-party to a two-party state. In seeking to break up the historic coalition between liberal Democrats and liberal Republicans, on the one hand, and conservative southern Democrats and conservative Republicans, on the other, Goldwater, Burns argued, wanted to leave the country with only two parties: one liberal and Democratic, the other conservative and Republican. Goldwater's response to such analysis was simply to say that he was seeking votes in the South because "you're supposed to hunt where the ducks are." Four years after Goldwater's 1964 debacle, Nixon nearly lost to Humphrey because the third-party candidate, George Wallace, drew so heavily in the South. Nixon's insight was that the South could be won by Republicans without playing the race card but instead by emphasizing the conservative, family-oriented, pro-law-and-order values northern Republicans had in common with white southern Democrats. The key was to work with the South on racial issues rather than seeking to disgrace Dixie and to support voting rights for all, thereby allowing blacks to flood into the Democratic Party across the South, accelerating the movement of conservative whites to the Republicans.

Buchanan notes that winning the South was not enough. Republicans in the North had to play the law-and-order card as well as other social issues in order to capture the white ethnic vote—those who, for example, supported Democratic mayors such as Chicago's Richard Daley and Philadelphia's Frank Rizzo. These constituencies were willing to accept an end to de jure segregation but rejected social experimentation, such as the busing of students to achieve numerical racial balance. They were more concerned about having their police drive criminals from the streets of their cities than they were about observing every conceivable procedural nicety imposed by the courts for the protection of suspected felons. They might have had some problems with the war in Vietnam, but they went into service when drafted. Often they volunteered. They had no patience for draft card destroyers, flag burners, and those who emigrated to Canada to beat the draft—let alone those like the Weathermen and Black Panthers who practiced violence to stop the war or achieve racial goals. Nixon recognized these people as the "great silent majority." He played to their opposition to lenient judges by pledging to appoint "strict constructionists" to the bench. He assured the South that its voice would be heard. Long before he took his first position on civil rights as president, Nixon's commitment to fairness under the law was said to be a casualty of his so-called "Southern Strategy."

Nixon's actual record on civil rights was far more progressive than his critics acknowledged. Early on he said that the approach of his administration would be to enforce court decrees in the civil rights area, rather than relying on his cabinet departments. But he was good to his word. When the Supreme Court decided *Holmes v. Alexander,* replacing the period of "all deliberate speed" with an "integrate now" command, Nixon, his vice president, Spiro Agnew, and, most notably, his extremely able secretary of labor, George Shultz, formed biracial committees of outstanding citizens in the affected states in order to ensure that the edict would be obeyed with minimal disturbance.

Nixon also supported extending the Voting Rights Act of 1965, described by many as the most important piece of civil rights legislation ever passed. He also advocated its extension to northern communities that met the act's low turnout criteria. As applied, the approach produced more elected blacks but fewer elected Democrats, particularly in the House of Representatives. Nixon harbored doubts about the innate abilities of African Americans to compete economically with whites. But by conviction, he was opposed to denying them that chance. His commitment to economic opportunity for minorities took many forms: automobile makers and other franchise industry leaders were urged to dispense more dealerships or outlets to minority businessmen. In the craft unions, many of which had been totally segregated, unions were compelled to adopt target numbers in order to bring blacks into parity with their numbers in the qualified population. That approach was tried first among Philadelphia craft unions, earning its place in history as the "Philadelphia Plan." The Office of Federal Contract Compliance (OFCC) also leaned on federal contractors to establish specific targets for the employment of black workers.

Nixon was also the first president to establish minority set-aside programs in the area of government contracts. Over time, these programs became mechanically administered and ran afoul of the Supreme Court. But in their day, thousands of minority contractors gained their first piece of federal action. And he supported giving teeth to the Equal Employment Opportunity Commission, which wrestled with cases of individual employment discrimination by backing judicial enforcement of its decrees.

With the help of his "closet Democrat," Daniel Patrick Moynihan, Nixon introduced the Family Assistance Plan (FAP), which essentially guaranteed a family of four an income of at least $1,600 per year. The plan, which also included welfare reform, was considered a beacon in its time. But representatives of the poor savaged the plan, in part to deny Nixon the luster of having achieved it and in part because they thought they might do better in some future administration. They're still waiting.

Nixon received little credit in his day for his backing of civil rights, and it is only recently that analysts such as Dean J. Kotlowski, whose book *Nixon's Civil Rights* offers a highly objective assessment of the Nixon record, are taking a second look. There are several reasons for this.

First, Nixon's was a nuanced program geared toward achieving an end to de jure segregation rather than forcing new kinds of experimental integration, such as mandatory busing or site selection policies designed to force minorities into predominantly white communities.

Second, Nixon's commitment to naming southern strict constructionists to the Supreme Court, as well as his support for tough new anticrime legislation, earned the distrust of minorities, who felt that in a crunch they would be abandoned to the president's "Southern Strategy." Nixon refused to demonize the South. His long-range objective was political—to bring it into the Republican Party purged of racism but not of conservatism.

Third, in the wake of Martin Luther King, Jr.'s death, the civil rights movement fractured between old liberals, such as Roy Wilkins, and radical new activists, such as Stokely Carmichael. The result was a division between black political leaders who saw their future organically linked to the Democratic Party and radicals who cared nothing for traditional political coalition building and sought instead to do their own racial thing. Both constituencies were hostile to Nixon-style republicanism. The leadership of the African-American community clearly saw its future as being linked to the Democratic Party. Franklin D. Roosevelt may have cracked the marriage between blacks and GOP dating back to the Reconstruction era, but it was not until the Nixon era that African Americans' support for the GOP bottomed out at under 10 percent. Evidence of Nixon's own personal bigotries has come to light over the years, particularly some of his more sordid anti-Semitic cracks. For a man whose immediate political family included the likes of William Safire, Henry Kissinger, Arthur Burns, Leonard Garment, and others, this has caused much embarrassment. "Nixon was not anti-Semitic," Ray Price insists. "But he did bitterly resent the almost universal antipathy Jews had for him."

An interesting exception that I noted time and again during the course of our interviews involved the people of Israel. Nixon would take almost any occasion to speak of their brilliance, their toughness, and their commitment with undisguised esteem. And when he recited accounts of policy disagreements, he left no doubt in my mind that he was as concerned with the best long-run interests of the Israelis as with the geopolitical dilemmas of his own government. In a sense, Nixon's feelings about Israel were as revealing as his obvious reverence for the Chinese leaders. For if the Chinese patriarchs had exhibited in their personal lives the ability to struggle, to fight, to lose, to come back, to fight again, to triumph over adversity, and to continue to struggle even after reaching the top, Israel's character as a nation was also rich in the sort of symbolism that Richard Nixon appreciated and revered.

No Harris Poll would show the Israelis unwilling to fight, and Nixon knew that no bearded, unkempt Israeli protestors would storm the Knesset urging peace at any price. The Israeli people did not seek haven as exiles when their country was in peril. Rather, they struggled to return home, to take up arms, to join the battle as citizen-soldiers.

Israeli leaders were not forced to seek fig leaves, they sought victory. Israeli victories were swift and certain, not agonizing and ambiguous. When Israel fought, there was no discussion of dominos. There was only one domino the Israelis cared about, and they knew it would fall the first time they lost a war.

Israel was smart and tough, two qualities that, when combined by Nixon, constituted the ultimate compliment he could bestow upon a nation or an individual. Israel would never be a superpower like the United States, but it would never be a pitiful helpless giant either.

Nixon's presidency saw major increases in the rights of women and came within one state legislature of adopting the Equal Rights Amendment, but Nixon's emotional and political involvement in the women's movement was negligible.

By contrast, Nixon's commitment to Native Americans' rights was perhaps the greatest of any president in history. He settled major land claims with the Taos Pueblo, the Yakama, and Alaska's native peoples; he ended talk of termination of the reservations; he reformed the archaic Bureau of Indian Affairs; and most Native Americans recognized that he was no "assimilationist," having been a process they equated with genocide.

For a man who barely mentioned the environment in his entire political life, Nixon is credited by many environmentalists as having had the best record of any administration in history. "He didn't start anything, but he let it all just happen," recently explained Douglas Starr, a specialist in environmental reporting affiliated with Boston University, "and what happened was the greatest proliferation of environmental legislation ever." Three factors contributed to the explosion of interest in environmental matters during the mid- to late 1960s: first, Rachel Carson's seminal book about pesticide pollution, *Silent Spring;* second, the 1968 oil spill in the Santa Barbara Channel; and third, Nixon's appointment of Governor Walter J. Hickel of Alaska as secretary of the interior at a moment when Hickel's biggest task would be to go thumbs-up or -down on construction of the Trans-Alaska Pipeline, a project many people feared would do irrevocable damage to the fragile Alaskan tundra or, worse yet, produce a devastating oil spill off the coast of Valdez, the proposed terminus.

Hickel presented a soft target. As an Alaskan office holder, he had never met a restriction on land use he could support, nor did he care much for isolated natural treasures. "A tree looking at a tree does nothing," he was fond of saying. Were there but a single tree in all of Alaska, environmentalists responded, Hickel would construct a sawmill on the site. Fortunately for their cause, the environmentalists decided to use Hickel's confirmation hearing as an educational forum for the dissemination of their concerns, rather than as a means of mobilizing support for his rejection. Nixon responded in kind, appointing the noted environmentalist Russell Train as Hickel's deputy and, shortly thereafter, forming the Council on Environmental Quality to dispense good advice to the president and his administration. Before his first term ended, Nixon would see the establishment of the Environmental Protection Agency and passage of historic clean air and clean water acts. "He wasn't mad about the environment," confessed his aide Ken Khachigian years later. "Nor did he care all that much about many domestic matters. His approach at the time was to give these people what they wanted in the hope that they would provide him with the support he needed to fight the war in Vietnam the way he knew it had to be fought in order to win."

Nixon's "new federalism" is portrayed by his supporters as an indication that his domestic views were not nearly as casual as critics maintained. True foreign issues were infinitely more important to him. But he was no less ahead of his time in calling for a return to the states of revenue sources then hoarded by the federal government along with the responsibility for spending the money wisely. More often than not the decision regarding expenditures belongs to those closest to the citizens. The problem with going very far in this analytical direction is that Nixon himself did not go very far with his new federalism. Concerned about creeping inflation, higher unemployment, and the growing general sense of malaise about the economy, Nixon elected to take the easy shot provided by Democrats and to impose wage and price controls after Congress passed legislation giving him the power to do so. Under Nixon, who imposed the controls essentially to distract attention from society's economic problems, the new federalism never had a real test.

Nixon did get to tinker with the Supreme Court of the United States. He appointed a total of four justices, more than any president save Franklin Roosevelt, who had problems of his own with the Court he inherited.

Nixon's first appointment, Chief Justice Warren E. Burger, proved to be a man of central casting looks but also of marginal intellectual and administrative talents. Harry Blackmun evolved into a dependable liberal during his years on the bench. William Rehnquist proved as able as he was conservative, while Lewis F. Powell, regarded as a somewhat aloof patrician at the time of his appointment, proved to be one of the most effective and influential members of the Court. Of course, two Nixon nominees—Clement F. Haynsworth and G. Harrold Carswell—failed to survive the nominating process. After berating the Senate for its anti-South bias following its rejection of Carswell, Mr. Nixon was forced to concede in our interviews that the appointment of the Florida mediocrity had been a mistake.

During and after the Nixon presidency, the Court, boasting up to four of his justices, attempted to abolish capital punishment, decreed that abortion came down to a woman's personal choice, permitted state universities to consider race a "plus factor" or "tie-breaker" in admission decisions, upheld the right of war dissenters to burn American flags, and in a unanimous decision told Mr. Nixon he must produce the Watergate tapes and other documentary evidence subpoenaed by the Office of the Special Prosecutor.

At the same time, while repealing none of the criminal decisions so hated by Nixon constituents, the Court clearly gave police and prosecutors some flexibility where search or arrest procedures had not been strictly adhered to or had not directly affected the outcome of the case.

Ironically, future generations may find the Court's decision in *United States v. Nixon* to be the most significant and far-lived "contribution" of the Nixon presidency. Decisions that deal with the allocation of power among the three branches of government often maintain their relevance over the centuries. Most constitutional scholars today can offer a lucid analysis of *Marbury v. Madison* (1803) or the 1819 case of *McCulloch v. Maryland*, but hardly any other decision from that time. In the same vein, *United States v. Nixon*'s role in limiting the scope of executive privilege may be on the lips of future constitutional scholars long after *Roe v. Wade* and *Regents of the University of California v. Bakke* are long-forgotten.

If *United States v. Nixon* will, through Nixon's own miscalculations, survive as his most enduring legacy, what about other decisions that seemed so important during his presidency or in the years immediately thereafter? Let us consider a few of the major candidates.

First, the new Republican Party.

I have discussed this new political alignment above, as well as Nixon's role in it. Nixon was among the most sophisticated politicians ever to seek higher office in the United States, and clearly he discerned the elements of what Kevin Phillips called "the emerging Republican majority." Nixon did what he could to encourage the process and speed it along. Clearly, he succeeded. But his contribution to what I believe was an inevitable process consisted of little more than saying some nice things about the South, holding hands with southern districts ordered to desegregate, and seeking to appoint a few southern judges to the U.S. Supreme Court. Yes, he also went after northern white ethnic voters, and yes, in 1972 and 1984 they voted Republican in mammoth numbers. But again, I think the Voting Rights Act of 1965 was the catalytic ingredient in realigning constituencies from four essential voting blocks into two. During the post-1965 period, the Sam Ervins, Senator Richard Russells, and the John Stennises were cast from the scene, to be replaced by the Jesse Helmses, the Bo Callaways, and the Trent Lotts. At the same time, northern liberal Republicans such as Jacob Javits, Hugh Scott, and Clifford Case were also disappearing. Nixon, for the most part, simply rode the crest of events. Nixon's courtship of the northern white ethnic vote is more impressive. With his tough anticrime stance, his appeals to patriotic values, his assaults on the media and other bastions of liberalism, Nixon put into play the political loyalties of hard hats and social conservatives, war veterans and police officers as had no other national Republican candidate in memory. That the opposition was dominated by the George McGovern left made that tactic even more successful. Nowhere was this reaching out more evident, the attempt to form an organic bond more clear, than in Mr. Nixon's "Silent Majority" address:

> *Let historians not record that when America was the most powerful nation in the world we passed on the other side of the road and allowed the last hopes for peace and freedom of millions of people to be suffocated by the force of totalitarianism.*
>
> *And so tonight—to you, the great silent majority of my fellow Americans—I ask for your support . . .*
>
> *Let us be united for peace. Let us also be united against defeat.*

Because let us understand: North Vietnam cannot defeat or humiliate the United States. Only Americans can do that.

After the passage of thirty-eight years since that address, the audience of former rock-solid Democrats is still one of the prized swing voter populations in U.S. politics, a true Nixon legacy, later afforded tender loving care by Ronald Reagan.

Second, détente with the Soviet Union.

This Nixon-Kissinger special barely outlived the 1972 trip organized to celebrate it. Ushered in at a time of manifest U.S. weakness in Vietnam, it was treated as a policy devised by accommodationists willing to live as the second-ranked power. The problem was compounded by widespread dissatisfaction with the arms control agreements themselves, outrage regarding the USSR's treatment of its dissidents and refuseniks, and the perception that the Russians had violated the very heart of the new arrangement by backing the Egyptian and Syrian Yom Kippur attacks against Israel and behaving even worse when Israel struck back successfully. Over the subsequent decade, détente was seen as a roadblock to successful U.S. action in either stemming Soviet acts of aggression or taking advantage of opportunities to hasten the demise of the brutal Communist system. Today, détente has no fathers, not Henry Kissinger, not Brent Scowcroft, not Dimitri Simes. It is a political orphan.

Third, the ABM Treaty.

This, together with SALT I, was a genuinely important achievement that preserved the ability of each superpower to deter the other from striking first in a crisis. Nixon was right, at the time the agreement was signed, in noting the folly of any power that would strike first, knowing that it could itself be destroyed by a retaliatory strike. But those who hated any deal implying acceptance of the permanence of the Soviet Union hit the ABM Treaty and accompanying restrictions on offensive missile launchers as a principal culprit of U.S. inferiority. The numbers were uneven, they declared. The deal could not prevent the Soviets from achieving a nuclear war–fighting capability. The United States was giving up a promising area—antimissile defensive systems—in which it had a clear technological advantage. Failing to deploy systems that could intercept nuclear weapons headed for one's population centers when the ability existed to do so was an act of total immorality. By the time the Soviet Union collapsed, even Nixon had joined the camp of his opponents and was marching in the parade for defensive missile deployments. On the grounds that a man should not prosper by disowning the very program he was most proud of, I would strike the ABM Treaty and accompanying limits on offensive missile launchers as being of marginal international consequence.

Fourth, Vietnam.

Richard Nixon and Henry Kissinger nearly pulled off a military and diplomatic miracle in Vietnam but wound up with nothing to show for it except Kissinger's Nobel Prize for negotiating a peace that died in its infancy. Those who criticized Nixon's handling of the war, his rallying domestic support against antiwar forces, and his occasional military surprises such as the incursions into Cambodia and Laos should have been more charitable at the time. He inherited a horrendous mess. He tried to end it by simultaneously using all three of the tools available to him: negotiation, linkage to steps he was taking with North Vietnam's Russian and Chinese allies, and unilateral withdrawal. Had they been attempted in succession, all would have failed miserably. That they were done simultaneously provided the two statesmen with at least a "Hail Mary" chance of success. Was it to be? No, because unilateral withdrawal or "Vietnamization" let the North Vietnamese know that it was not necessary to negotiate in order to get the Americans out of

South Vietnam. Linkage couldn't work because both the Soviets and Chinese knew that the United States was far more committed to developing serious relationships with themselves than it was to reaching a satisfactory conclusion in Vietnam. In the end Nixon got the best deal he could but still left his South Vietnamese ally in a precarious position, given that the deal permitted North Vietnamese and Viet Cong forces in the South to remain in place. During the period of the "imperial presidency," Nixon and Kissinger might just have gotten a commitment from Congress to enforce the deal. But that was so far from what the Congress of 1973–1975 was willing to commit to that it strains credulity to think that Nixon and Kissinger thought they might pull it off.

Fifth, China.

This was a great step on Nixon's part, propitiously timed, enthusiastically pursued, professionally executed. There were minefields along the path to normalization—détente with the Soviets, the Taiwan question, a war in Vietnam—but the administration skirted them all. Nixon did not go to China in order to change the society Mao Zedong was about to leave behind, although change it has during the subsequent thirty-five years. Nor did he go there to force consideration of the Taiwan question—that, too, has been successfully finessed over time. He did seek help in Vietnam but understood he could not subordinate other goals to that evasive objective. So he signed a final communiqué, worked out before he arrived, expressing accord on those matters in which the two sides could find agreement and soft-pedaling other areas that, if pushed, could have ruined the entire event.

Nixon's old friends in the China Lobby did feel betrayed. Chinese political writer Anna Chennault and others spread reports that Nixon had tricked them by making all sorts of promises of support during his campaign if the lobby would publicly oppose Lyndon Johnson's bombing halt in North Vietnam. Other complications loomed down the road, including the Taiwan question, China's regional intentions, and perhaps even China's desire to become the world's next superpower. There may, in short, come a time when scholars and historians critique Nixon's China overture as the key that unlocked the door to an era when the United States would be overshadowed by the world's most populous and fastest-developing nation. But from the admittedly constrained perspective of thirty-five years, Nixon—the one man with the credentials to flip China policy—did a right and profoundly important thing.

Sixth, Watergate.

There is no question with which I have grappled more since my interview with Nixon than that of where Watergate fits into the perspective of history. For purposes of approaching this analysis, I have taken the position that what has not been proved by this time should be considered disproved. Therefore, I take at face value Mr. Nixon's assertion that he knew nothing in advance of the Watergate break-in and that his essential motivation in directing Haldeman to lean on the CIA in order to block the FBI investigation was undertaken for purposes of minimizing political fallout from the incident. That, of course, does not absolve Mr. Nixon of the crime of obstructing justice. Indeed, given the statutory language involved, it constitutes an admission of guilt.

Watergate is a funny animal. Planting a bug on the phone of your chief political rival is not the worst sin on Earth, though it may well be a felony. The remaining offenses—suborning perjury, paying for the silence of witnesses, firing an honest official during the Saturday Night Massacre—get worse and worse but would probably not get my juices flowing to the extent that Watergate did had not most of the culprits been members of the U.S. government during the Nixon presidency. Honest governments are not that easy to find. Many systems the world over are predicated on the use of power, ballot box stuffing, and fraud. Democracies are different. They operate by the consent of the governed. They stand as a pyramid of good faith, of the honest execution of the law, of a respect for the opinions of one's fellow citizens. The risk that a less qualified candidate will reach office, that the less qualified party will dominate the legislature, is subordinated to the risk taught by experience again and again that the dangers of authoritarian or totalitarian systems are far more egregious. When elections are dishonest, democracy doesn't work. When processes are suborned, the system breaks down. When all this is done by people not in the government, it is bad enough. When it is done by subcabinet appointees or even those of cabinet rank, it is even worse. When it is done by the president of the United States, it is a shocking crime, a crime against the system, a crime against its values, a crime against its people. That is what Richard Nixon and his Watergate crimes were all about. What was at stake in the lengthy and diverse Watergate investigation was the question of whether this democracy could purge itself, could restore its institutions to robust health, could restore an increasingly cynical electorate to good citizenry, could restore society's basic confidence in the institutions created to reflect it.

Though society's willingness to restore Nixon to elder statesman status was some reflection of its desire to move beyond Watergate, the public cynicism engendered by those crimes seems destined to repeat itself every few years as one "gate" or another—"Lancegate," "Koreagate," "Iraqgate," "Iran-Contra-gate," "Monicagate"—commands the public attention. But Watergate was unique.

I don't believe, after twenty-eight and three quarters hours of interviewing the man, that Richard Nixon or his key defenders ever absorbed that message. Richard Nixon said he let the country down, but my feeling was that he was even more remorseful for letting himself down by getting caught. He might have escaped unscathed by burning the tapes. He could have shut his door and his ears to all that was going on about him. He could have fired all the culprits, including himself, and pardoned them to keep them from going to prison. He could have stalled the process through a battle in the Senate, hoping the numbers would stop accumulating short of the two-thirds majority needed to expel him from office. Instead, "I gave them a sword and they stuck it in, and they twisted it with relish, and if I'd been in their position I'd have done the same thing." A tactical mistake. A game lost. A careless maneuver on the field of battle. A loss instead of a win. Nothing more at issue than the identity of winner and loser; possibly the news should have been carried in the sports section rather than on the front pages.

I truly believe that Richard Nixon went to his grave firmly believing that he was a victim. Not in the sense that he didn't run afoul of the law or perform the acts necessary to incriminate himself. Rather that he was a victim of a media establishment that hated him, left-wing Democrats still sulking over the outcome of the Jerry Voorhis contest, academics and intellectuals who came to believe Alger Hiss was guilty but who could never forgive Richard Nixon for proving it. Nixon in many ways was a supremely talented man. To praise his analytical skills, his great experience in foreign policy, his empathy for the values of the silent majority is not to simply throw a sop his way. Very few presidents of the past century came close to Nixon's skills in foreign affairs or his ability to empathize with ordinary people. But to dismiss his crimes by simply noting that the man had a dark side is not quite sufficient. This was a man whose dark side conquered the whole. This was a man with competing tendencies that could not coexist. He is not a man who threatened democratic society in the manner of, say Augusto Pinochet or Hugo Chávez; rather, he threatened it with corrosion from within, with a lack of structural integrity, with destruction of the value system upon which it is grounded.

The pardon by Gerald Ford seemed a relatively small and somewhat technical act. Nixon's real pardon would have to come from a higher source.

6

TAKING MY LEAVE OF RICHARD NIXON, MAY 1977

I saw Richard Nixon only once more. It was a little over a week later, just after the second program on foreign policy had been broadcast. The reaction to the interviews by both press and public had been greater than I could have dreamt possible. John and I had completed the editing of the third and fourth programs, and I was leaving California. Before I left, I drove to San Clemente with Caroline Cushing to say farewell to Richard Nixon.

"Hello, Mr. President," I begin.

"Hello . . . David." It is an affectionate greeting in its way. Never before has he called me by my first name. Four weeks ago I would have predicted that, once the post-Easter interrogation was over, we would have no communication between us, not even the pretense of a relationship. Yet Nixon told his friends that he regarded the interviews as "tough but fair," two words that one would have said were mutually exclusive in his vocabulary. He is full of surprises.

The opinion polls and the press, which have been so positive about the program, have not been as positive about him. I feel I should murmur my condolences. None is in order. He had expected nothing better from the media. "You knew they'd crap on that, didn't you?" he asks. The mail coming to San Clemente has been pretty good. And many of his friends and former colleagues have told him he did much to purge the poison of Watergate from his own system, and perhaps the country's.

There are globes in Richard Nixon's house and globes in his office. He looks at them often, studies them, cradles them with his hands as a gypsy fortune-teller does her crystal ball. There is a certain mysticism about it all. It seems the closest he comes to the formal practice of religion.

I look at Richard Nixon, and I see the face of tragedy. He is an intelligent man, in many ways an incredibly able man. He thinks clearly, and he speaks well. He is a man to whom history has relevance. He has a sophisticated understanding of world affairs, a nice touch for dealing with other leaders. He might have made a good secretary of state. Perhaps a great one.

Yet of all the strengths he has talked about, the one he has ignored strikes me as the most critical. And that is the strength that comes from a nation's belief in the essential rightness of its own cause, in the integrity of its own vision, in the justice of its own ends, in the basic goodness of its own deeds. And that is the strength that was undermined, the faith that was shaken, by Vietnam abroad and by Watergate at home. When Richard Nixon talks so sincerely, so articulately, of the developments that he fears and deplores—a reluctance on the part of the United States to assume its global responsibilities as he defines them and a weakened presidency that makes it more difficult to shake the nation out of its lethargy—he is talking, in part at least, of his own legacy.

But that part of our conversation is over. And once again he confounds the caricatures of himself. He takes Caroline by the hand, firmly, warmly—this man supposedly so uncomfortable with women. He takes her to the window. "Out there is China," he tells her, pointing and gazing in a dreamy way. "Let me show you the garden . . ." and he leads her out onto the patio. "Brezhnev slept in that room," he tells her. "A great swordsman. The Russians are, you know. Have you read Tolstoy? *Anna Karenina,* very romantic . . ."

They return. For a moment, the usually somber cloud has lifted. Manolo Sanchez brings some more blanc de blancs. "Get the caviar the shah sent us for Christmas, Manolo." Manolo is requested to do his favorite imitation: of Henry Kissinger biting his nails, clutching his files, and losing his toothbrush. There is a moment of genuine gaiety. And then the spell is broken. The more somber tone returns. But for a moment it was there. And I had not expected it. Any more than four weeks ago I would have expected him to touch me as he had in our Watergate sessions. To break down, as he had, the barriers to intimacy that he had erected so painstakingly through the years.

But we have trespassed upon his solitude for long enough. It is time to go. We leave him standing by the window, gazing toward the ocean. He has made us feel at home. This man normally so ill at ease with people. Perhaps even more ill at ease with himself. A good mind, with a thirst for nobility. A sad man, who so wanted to be great.

As we drive away, I look back and I wish him peace at the center.

Part II

TRANSCRIPTS

7

WATERGATE

We knew that Watergate would be the key moment of the Frost/Nixon interviews. If we failed every other subject but scored impressive points on Watergate, the interviews would be successful. If we did every other thing right but got Watergate wrong, we would fail. The strategy was to confront Nixon at every turn. If his factual statements were wrong, challenge them. If his interpretation of the law was off base, correct it. Know the record as thoroughly as it could be known, and let him know that we know it so he will abandon attempts to run roughshod over us. Most important, never stray from our theory of the case. The cover-up began within hours of the Watergate arrests. The president was apprised about those involved early on. He orchestrated efforts to impede and divert those investigating the crime while buying the silence of those who had committed it.

FROST: Mr. President, to try and review your account of Watergate, ah . . . in one program is a daunting task, but, ah, we'll press first of all through the sort of factual record and the sequence of events as concisely as we can to begin with. But just one brief, preliminary question: Reviewing now your conduct over the whole of the Watergate period, with the additional perspective now of three years out of office and so on, do you feel that you ever obstructed justice? Or were part of a conspiracy to obstruct justice?

NIXON: Well, in answer to that question, I think that the best procedure would be for us to do exactly what you're going to do on this program; ah, to go through the whole record in which I will, ah, say what I did; ah, what my motives were; ah, and, ah, then I will give you my evaluation as to, ah, whether those actions or, ah, anything I said, for that matter, ah, amounted to what you have called an obstruction, ah, of justice. Ah, I will express an opinion on it, but I think what we should do is to go over it, ah, the whole matter, so that, ah, our viewers will have an opportunity to know what we are talking about. Ah, ah, so that in effect, ah, they, as they listen, ah, will be able to hear the facts, ah, make up their own minds. I'll express my own opinion. They may have a different opinion. You may have a different opinion. Ah, but that is really the best way to do it, rather than to preclude it in advance and maybe prejudice their viewpoint.

FROST: I'm very happy to do that, because I think the only way, really, to examine all of these events is on a blow-by-blow account of, of what occurred. So, beginning with June 20, then, what did Haldeman tell you during the eighteen-and-a-half-minute gap?

NIXON: Haldeman's notes, ah, are the only recollection I have of what he told me. Haldeman was a very good note taker, ah, because, of course, we've had other opportunities to look at his notes and he was very . . . he was making the notes for my presidential files. The notes indicated—

FROST: PR offensive and—

NIXON: That's right.

FROST: —all of that.

NIXON: Well, of course. Ah, they . . . the notes were—

FROST: Diversion.

NIXON: Well, you've asked me what it was. My recollection was that the notes showed . . . "Check the EOB to see whether or not it's bugged." Obviously, I was concerned about whether or not the other side was bugging us. I went on to say, ah, "Let's get a public relations offensive on what the other side is doing in this area and so forth," ah, and in effect, ah, ah, "Don't allow, ah, the, ah, ah, Democratic opposition, ah . . . to build this up, ah, into basically, ah . . . blow it up into a big political issue." Those were the concerns expressed. And I have no recollection of the conversation except that.

Our first intimation of Nixon's defense to involvement in the cover-up came as we confronted him with a previously unreported tape of a June 20, 1972, conversation between the president and his political aide Charles Colson. In discussing the break-in with Colson, the president used words consistent with bottling up the facts. How could he explain this early planning to keep significant facts from coming to light?

FROST: But as far as your general state of knowledge that evening, ah, when you were talking with Chuck Colson on the evening of June the twentieth, it suggests that, from somewhere, your knowledge has gone much further. You say, "If we didn't know better, we'd have thought the whole thing had been deliberately botched." Colson tells you, "Bob is pulling it all together. Thus far, I think we've done the right things to date." And you say, "Ba . . . ah, basic . . ." He says, "Basically, they're all pretty hard-line guys." And you say, "You mean, Hunt?" And you say, "Of course, we're just gonna leave this where it is with the Cubans. At times, I just stonewall it." And you also say, "We gotta have lawyers smart enough to have our people delay." Now, somewhere you were pretty well informed by that conversation on June 20.

NIXON: As far as my information on June 20 is concerned, ah, I had been informed ah, by . . . with regard to the possibility of Hunt's involvement, ah, whether I knew on the twentieth or the twenty-first or twenty-second, I knew something . . . I learned in that period about the possibility of Liddy's involvement. Of course, I knew about the Cubans and McCord, who were all picked up at the scene of the crime. Ah, no, ah, you have read here, ah, excerpts out of a conversation with Colson. Ah, and, let me say, as far as my motive was concerned—and that's the important thing—my motive was, in everything I was saying or certainly thinking at the time, ah, ah, was not, ah, to try to cover up a criminal action. But to be sure that, as far as any slip-over, or should I say "slop-over," I think would be a better word, any slop-over in a way that would, ah, ah, damage innocent people or blow it into political proportions . . . it was that that I certainly wanted to avoid.

FROST: So you invented the CIA thing on the twenty-third as a cover?

Once again Nixon will incorrectly offer the purity of his motives as a defense of conduct that is clearly criminal. Even at this early stage of the interrogation it was possible to get a read on Nixon's approach. He was tough-minded, self-confident, and unflinching in his own defense. He was giving no ground. He was testing my knowledge of the law and the various transactions that had led to the cover-up charges. He would yield only when members of his own staff warned him that the hard line he was taking would impede his efforts to again become a member in good standing of the national political community.

NIXON: No. Now, let's . . . let's use the word *cover-up,* though, in the sense that it had . . . should be used and should not be used. If a cover-up is for the purpose of covering up criminal activities, it is illegal. If, however, a cover-up, as you have called it, is for a motive that is not criminal, that is something else again. And my motive was not criminal. I didn't believe that we were covering any criminal activities. Ah, I didn't believe that John Mitchell was involved. Ah, I didn't believe, ah, that, ah, for that matter, anybody else was. I was trying to contain it politically. And, that's a very different motive from the motive of attempting to cover up criminal activities of an individual. And so there was no cover-up of any criminal activities; that was not my motive.

FROST: But surely, in all you've said, you have proved, exactly, that that was the case; that there was a cover-up of criminal activity because you've already said, and the record shows, that you knew that Hunt and Liddy were involved; you'd been told that Hunt and Liddy were involved. At the moment when you told the CIA to tell the FBI to "stop, period" as you put it. At that point, only five people had been arrested. Liddy was not even under suspicion, and so you knew in terms of intent, and you knew in terms of foreseeable consequence, ah, that the result would be that, in fact, criminals would be protected. Hunt and Liddy, who were criminally liable, would be protected. You knew about them. The whole statement says that, ah, ah, "We . . . we're gonna . . ." Haldeman says, "We don't want you to go any further on it. Get them to stop. They don't need to pursue it, they've already got their case." Walter's notes, that he said, "Five suspects have been arrested, this should be sufficient." You said, "Tell them, don't go any further into this case. Period." By definition, by what you've said and by what the record shows, that, per se, was a conspiracy to obstruct justice because you were limiting it to five people when, even if we grant the point that you weren't sure about Mitchell, you already knew about Hunt and Liddy and had talked about both, so that is obstruction of justice—

NIXON: Now, just a moment.

FROST: —period.

NIXON: Ah, that's your conclusion.

FROST: It is.

NIXON: But now let's look at the facts. Ah, the fact is that, as far as Liddy was concerned, ah, what I knew was . . . was only the fact that, ah, he was the man on the committee, ah, who was in charge of intelligence operations. As far as Hunt is . . . was concerned, ah, and if you read that tape, you will find I told them to "tell the FBI"—they didn't know, apparently—"and the CIA that Hunt was involved." And so there wasn't any . . . any attempt, ah, to, ah, keep them from knowing that Hunt was involved. The other important point to bear in mind when you ask, "What happened?" and so forth is what happened two weeks later. Ah, two weeks later when, ah, I was here in San Clemente, I called Pat Gray, the then FBI director, on the phone to congratulate the FBI on a very successful operation they'd had in apprehending some hijackers in San Francisco or someplace abroad. He then brought up the subject, ah, of the Watergate investigation. He said, "There are some people around you who are mortally wounding you, or would . . . might mortally wound you because they're trying to restrict this investigation." And I said, "Well, have you talked to Walters about this matter?" And, he said, "Yes." I said, "Does he agree?" He said, "Yes." I said, "Well, Pat"—I know him . . . had known him very well, of course, over the years, I did call him by his first name. I said, "Pat, you go right ahead with your investigation." He has so testified, and he did go ahead with the investigation.

Nixon used this July 6 conversation with acting FBI Director L. Patrick Gray to purge himself of any culpability for the criminal cover-up. After all, when Gray told him that aides were trying to "mortally wound" Mr. Nixon, the president told him to go on with his investigation. The problem with this defense is twofold: first, telling the director to conduct a full investigation while simultaneously orchestrating a cover-up is hardly the stuff of innocent behavior. Second, as I would quickly point out, Nixon had already been participating in a cover-up at least since June 23, and probably before that.

FROST: Yes, but the point is that, ah, obstruction of justice is obstruction of justice if it's for a minute or five minutes, much less for the period June 23 to July the fifth, when I think it was when he talked to Walters and decided to go ahead, the day before he spoke to you on July the sixth. It's obstruction of justice how . . . for however long a period, isn't it? And, also, it's no defense to say that the plan failed; that the CIA didn't go along with it, refused to go along with it, that it was transparent. I mean, if I try and rob a bank and fail, that's no defense. I still tried to rob a bank. I would say you still tried to obstruct justice and succeeded for that period. He's testified they didn't interview Ogarrio—

NIXON: Now, let's—

FROST: —they didn't do all of this; and so I would have said it was a successful attempt to obstruct justice for that brief period.

NIXON: Now, just a moment, ah, you're again making the case, which of course is your responsibility as the attorney for the prosecution, ah, let me make the case as it should be made, ah, even if I were not the one, ah, who was involved ah, for the defense. The case for the defense here is this: you use the term "obstruction of justice." You perhaps have not read the statute with regard to respect . . . ah, ah, ah, obstruction of justice—

FROST: Well, I have.

NIXON: Obstruction . . . well, oh, I'm sorry, of course, you probably have read it, but possibly you might have missed it because when I read it, ah, many years ago, in, ah, ah, perhaps when I was studying law, although the statute didn't even exist then, because it's a relatively new statute, as you know. But in any event, ah, when I read it even in recent times, ah, I was not familiar with all of the implications of it. The statue doesn't require just an act. The statute has the specific provision: one must corruptly impede a judicial—

FROST: Well, you . . . a corrupt—

NIXON: —matter.

FROST: —endeavor is enough.

NIXON: A con . . . con . . . all right, we'll . . . a conduct . . . endeavor. Corrupt intent. But it must be corrupt, and that gets to the point of motive. One must have a corrupt motive. Now, I did not have a corrupt motive.

FROST: You . . . you were—

NIXON: My motive was pure political containment. And political containment is not a corrupt motive. If so, for example, we . . . President Truman would have been impeached.

FROST: But the point is that . . . the point is that your motive can be helpful when intent is not clear. Your intent is absolutely clear; it's stated again, "Stop this investigation here. Period." The foreseeable, inevitable consequence, if you'd been successful, would have been that Hunt and Liddy would not have been brought to justice. How can that not be a conspiracy to obstruct justice?

NIXON: No. Wait a minute. "Stop the—"

FROST: You would have protected—

NIXON: "Stop this—"

FROST: —Hunt and Liddy from guilt.

NIXON: "Stop the investigation." Ah, eh, you still have to get back to the point that I have made, ah, previously, that, when I . . . that, ah, that my concern there, which was conveyed to them, and the decision then was in their hands. Ah, my concern was having the investigation spread further than it needed to.

FROST: Well . . .

NIXON: And, as far as that was concerned, ah, I don't believe, as I said, we turned over the fact that we knew that Hunt was involved, that a possibility that Liddy was involved, ah, but under the circumstances—

FROST: You didn't turn that over, though, did you?

NIXON: What?

FROST: You didn't turn that over.

NIXON: No, no, no, no, no. We turned over the fact that Hunt, that, that Hunt was . . . was involved.

FROST: You never told anyone about Liddy, though.

NIXON: No, not at that point.

FROST: Now after the Gray, ah . . . after the Gray conversation, the cover-up went on. You would say, I think, that you were not aware of it. I was arguing that you were a part of it as a result of the June the twenty-third, ah, conversations. But you would say that you were—

NIXON: Are you sure I was a part of it as a result of the June 23 conversations?

FROST: Yes.

NIXON: Ah, after July 6, when I talked to Gray?

FROST: I would have said that you joined the conspiracy, which you therefore never left.

NIXON: Yes, no. Well, then we totally disagree on that.

FROST: But, I mean . . . those are the two positions.

NIXON: That's right.

FROST: Now, you, in fact, however, would say that you first learned of the cover-up on March the twenty-first. Is that right?

NIXON: On March 21 . . . was the date when I was first informed of the fact, the important fact to me in that conversation, ah, was of the blackmail threat that was being made by Howard Hunt, who was one of the Watergate, ah, ah, participants, but not about Watergate.

At one point in his career, CIA agent Hunt was involved in many sensitive security matters. But his White House activities seemed to be mainly of the dirty tricks variety. He spirited the ITT lobbyist Dita Beard out of town when an embarrassing memo she had penned came to light. He scurried to Chappaquiddick to gather derogatory material on Ted Kennedy after his fatal accident. He was involved in the break-in into the office of Dr. Lewis Fielding, Daniel Ellsberg's psychiatrist. And his White House safe contained forged cables purporting to show President John F. Kennedy's approval of plans by South Vietnamese army officers to murder President Ngo Dinh Diem. One can imagine Mr. Nixon not wanting these activities to come to light, but protecting them by orchestrating a criminal cover-up seems only to compound the crime.

FROST: So, during the period between those two dates, between the end of June, beginning of July, and March the twenty-first, ah, while lots of elements of the cover-up, as we now know, were continuing, were you ever made aware of any of them?

NIXON: No. I . . . I don't know what you're referring to.

FROST: Well, for instance, your personal lawyer, Herbert Kalmbach, coming to Washington to start the raising of, ah, $219,000 of hush money, approved by Haldeman and Ehrlichman. They went ahead but without . . . without clearing it with you?

People who invent innocent motives to explain illegal transactions sometimes sound funny to the point of absurdity. Nixon certainly fit this paradigm as we discussed the raising and distribution of hush money for the Watergate defendants.

NIXON: That was one of the statements that I've made, ah, which, ah, after all of the checking we can possibly do . . . we checked with Haldeman, we checked with Ehrlichman. I wondered, for example, if I had been informed. If I had been informed that money was being raised for humanitarian purposes, to help these people with their defense, I would certainly have approved it. If I had been told that the purpose of the money was to raise it for the purpose of keeping 'em quiet, I would have been . . . disapproved it.

FROST: But—

NIXON: But the truth of the matter is that I was not told. I did not learn of it until the March period.

FROST: But in that case, if that was the first occasion, why did you say in, ah, such strong terms to Colson, on . . . on February the fourteenth, which is more than a month before, you said to him, "The cover-up is the main ingredient; that's where we gotta cut our losses; my losses are to be cut; the president's loss has gotta be cut on the cover-up deal."

NIXON: When did I say that?

FROST: February the fourteenth.

NIXON: Well, because I read the American papers. And in January, the stories that came out, they're not . . . not just from *The Washington Post*—the famous series by some unnamed correspondents who have written a best-selling book since then—ah, but, *The New York Times*, the networks, and so forth, were talking about "hush money." They were talking about clemency pay . . . ah, ah, for cover-up, and all the rest. It was that that I was referring to at that point. I was referring to the fact that there was a lot of talk about cover-up and that this must be avoided at all cost.

FROST: But, there's one, ah, very clear, self-contained quote, and I read the whole of this conversation of February 14, which I don't think has ever been published, but . . . and there was one very clear quote in it that I thought was—

NIXON: It hasn't been published, you say?

FROST: No, I think it's . . . it's available to anybody who consults the records, but, ah—

NIXON: Oh. Yes.

FROST: —but, ah, people don't consult all the records.

NIXON: Just wondered if we'd seen it.

FROST: Well, I'm . . . I'm sure you have, yes, but ah . . . where the president says this, on February the thirteenth, um, "When I'm speaking about Wa—"—this is to Colson—"When I'm speaking about Watergate, though, that's the whole point of the election. This tremendous investigation rests unless one of the seven begins to talk, that's the problem." Now, in that remark, it seems to me that someone running the cover-up couldn't have expressed it more clearly than that, could they?

Once again Nixon works familiar territory in explaining the noncriminal motives of the criminal cover-up. But this time he adds a new element by claiming he had adopted the role of the attorney for the defense and that his suggestions of how to act and what to say had been intended not to encourage further illegal conduct but rather to explain his dealing to date in the most favorable light.

NIXON: What . . . what do we mean by "one of the seven beginning to talk"? I've . . . how many times do I have to tell you? Ah, that as far as these seven were concerned, ah, the concern that we had, certainly that I had, ah, was that men, ah, who, ah, worked in this kind of a covert activity, men who, of course, ah, realize it's dangerous activity to work in, particularly since it involves illegal entry, ah, that, ah, once they're apprehended, ah, they are likely to say anything. And the question was, I didn't know of anybody at that point, nobody on the White House staff, not John Mitchell, anybody else that I believed, ah, was involved . . . ah, criminally. Ah, but on the other hand, I certainly could . . . could believe that a man like Howard Hunt, who was a prolific book writer, or any one of the others under the pressures of the moment, ah, could have started blowing and putting out all sorts of stories, ah, to embarrass the administration. And, as it later turned out in Hunt's case, to blackmail the president to provide clemency or to provide money or both.

FROST: I still just think, though, that one has to go contrary to the normal usage of language of almost ten thousand gangster movies, ah, to interpret "This tremendous investigation rests unless one of the seven begins to talk, that's the problem" as anything other than some sort of conspiracy to stop him talking about something damaging—

NIXON: Well, you can . . . you can state—

FROST: —to the press and making the speech.

NIXON: —you can state your conclusion, and I've stated my views.

FROST: That's fair.

NIXON: So now we go on with the rest of it.

In one of the most pivotal Watergate conversations, Nixon and Dean met on March 21, 1973, at which time Dean informed the president that unless more money was forthcoming the defendants might "blow." Of course, by this time the cover-up was in such a state of havoc that its participants, including the president, were grappling with multidimensional requirements: explain your actions; try to skirt the clear meanings of words; scuttle those staffers too deeply involved to save; deal with a White House counsel who had so despaired of saving the presidency that he was now in business strictly for himself.

FROST: Looking back on the record, now, of that conversation—as I'm sure you've done in addition to the overall details, which we'll come to in a minute—bearing in mind that a payment probably was set in motion prior to the meeting and was certainly not completed until late the evening of the meeting, wouldn't you say that the record of the meeting does show that you endorsed or ratified what was going on, with regard to payment to Hunt?

NIXON: No, the record doesn't show that at all. In fact, ah, the record actually is ambiguous, ah, until you get to the end, and then it's quite clear. And what I said . . . the . . . later in the day, and what I said the following day, shows what, ah, the facts really are and completely contradicts the fact . . . the point that has been made, and, ah, again here's a case where Mr. Jaworski, in his book, conveniently overlooks, ah, what actually was done and what I did say the following day, ah, as well as, ah, ah, other aspects of it. Let me say, I did consider the payment of $120,000 to

Hunt's lawyer and to Hunt, ah, for his attorney's fees and for support. Ah, I considered it not because Hunt was gonna blow, using our gangster language here, on Watergate because, as the record clearly shows, Dean says, ah, "It isn't about Watergate, but it's going to talk about some of the things he's done for Ehrlichman." Ah, as far as the payment of the money was concerned, when the total record is read, you will find that it seems to end on a basis which is indecisive, ah, but I clearly remember, and you undoubtedly have it in your notes there, ah, my saying that "The White House can't do it." I think for my . . . were my last words. Ah, because I had gone through the whole, ah, scenario with, ah, Dean and I laid it out, I said, "Look, what would it co— I mean, when you're talking about all of these people, what would it cost to, ah, take care of them for—"

FROST: Well, no, I'm . . . I've—

NIXON: —and we talked about a million dollars. And I said, "Well, you could raise the money, but doesn't it finally get down to a question of clemency?" And he said, "Yes." I said, "Well, you can't provide clemency, and that would be wrong for sure." Now, if clemency's the bottom line, then providing money isn't going to make any sense.

We now come to what was perhaps the single most significant moment of the Nixon interviews. In preparing to interrogate Nixon, I had assembled a list of some of the most damning references for the need to continue blackmail payments to Howard Hunt and the other Watergate defendants. Most were drawn from the president's March 21 conversation with John Dean. As I read the various quotes, Nixon's demeanor changed. Gone were the expression of confidence, the resolve to steamroller his adversary, the flimsy ex-

cuses that followed flimsy excuses over hours of interrogation. As I read the quotes, Nixon's face became drawn and strained; each quote somehow seemed to have the impact of a blow on the ropes of a virtual boxing ring. Those who today observe the tape suggest that at this moment Nixon knew he was a beaten man. Clearly something had struck home.

FROST: But when you . . . we talk about the money, the $120,000 demand that, in fact, he got $75,000 of that evening, bearing in mind what you were saying earlier about, reading that the overall context of the conversation, is there any doubt, when one reads . . . reading the whole conversation:

1. "You could get a million dollars, and you could get it in cash. I know where it could be gotten."
2. "Your major guy to keep under control is Hunt?"
3. "Don't you have to handle Hunt's financial situation?"
4. "Let me put it frankly: I wonder if that doesn't have to be continued."
5. "Get the million bucks, it would seem to me that would be worthwhile."
6. "Don't you agree that you'd better get the Hunt thing?"
7. "That's worth it, and that's buying time."
8. "We should buy the time on that, as I pointed out to John."
9. "Hunt has at least got to know this before he's sentenced."
10. "First, you've got the Hunt problem, that ought to be handled first."
11. "The money can be provided. Mitchell could provide the way to deliver it. That could be done. See what I mean?"

12. "But let's come back to the money. [They were off on something else here, desperate to get away from the money, bored to death with the continual references to the money.] A million dollars and so forth and so on. Let me say that I think you could get that in cash."
13. "That's why your immediate thing . . . you've got no choice with Hunt but 120 or whatever it is. Right?"
14. "Would you agree that this is a buy-time thing? You'd better damn well get that done . . . but fast."
15. "Now, who's gonna talk to him? Colson?"
16. "We have no choice."

And so on. Now, reading as you've requested—

NIXON: All right, fine.

FROST: —within the whole context, that is—

NIXON: Let me, let me just stop you right there. Right there. You're doing something here which I am not doing and I will not do throughout these broadcasts. You have every right to. Ah, you were reading there out of context, ah, out of order, because I have read this and I know—

FROST: Oh, I know.

NIXON: —it really better than you do.

FROST: I'm sure you do.

NIXON: And, and I should know it better because I was there. It's no reflection on you. You know it better than anybody else I know, incidentally, and, ah, you're doing it very well. But I am not going to sit here and read the thing back to you. I could have notes here; as you know, I've participated on these broadcasts without a note in front of me. I've done it all from recollection. I may have made some mistakes.

FROST: No, you—

NIXON: But not many.

FROST: I . . . you, you certainly have done that—

Nixon tried one more defense: he might have ruminated about paying Hunt, but Hunt wanted something the president could not afford to give him, and that was clemency. So, reasoned Nixon, if clemency was not a viable option, than neither was hush money alone, and there was no reason for him to have authorized it.

NIXON: Now, let me say this, and let me say—

FROST: —and I agree with you, it's your life we're talking about.

NIXON: —that in this instance, that in this instance, the very last thing you read, ah, "Do you ever have any choice with Hunt?" It . . . why didn't you read the next sentence? Why did you leave it off?

FROST: It carried on.

NIXON: No, no. The reason . . . the next sentence says, as I remember that so well, "But ya never have a choice with Hunt. Do you ever have one?" Rhetorically, you never have a choice with Hunt. Because when you finally come down to it, it gets down to clemency. Now, why after all of that horror story? And, it was, I mean, even considering that, I mean, must horrify people. Why would you consider paying money to somebody who's blackmailing the White House? I've tried to give you my reasons. I was concerned about what he would do. But my point is: After that, why not? Why not do what was not done by Mr. Jaworski in his book? What was not done by Mr. Doar before the Senate Judiciary Committee? Read the last sentence. The last sentence, which says after that, "You never have any choice with Hunt, because it finally comes down to clemency." And I said six times in that conversation, you didn't read that in your ten things, six times I said, "You can't provide clemency."

FROST: No, I said—

NIXON: "It's wrong for sure."

FROST: No, I never said there . . . I never said there that you did provide clemency, nor was I talking about—

NIXON: My point is—

FROST: —but I was—

NIXON: My point is—

FROST: —all right, let me quote—

NIXON: My point is that without—

FROST: —let me quote to you, then, I've been through the record and I want to be totally fair, let me read to you the last quote on the transcripts that I can find about this matter then. You said, "Why didn't I go to the last one?" I read sixteen, and I thought that was enough but . . . we could have read many more than that. But the last thing in the transcripts I can find about this subject was you talking on April the twentieth, and you were recollecting this meeting and you said that you said to Dean and to Haldeman, "Christ, turn over any cash we got." That's your recollection of the meeting on April the twentieth, when you didn't know you were on television.

NIXON: Of course I didn't know I was on television. On April the twentieth, it would well have been my recollection. But my point is: I wonder why, again, we haven't followed up with what happened after the meeting. Let me tell you what happened after the meeting. And, and you were, incidentally, very fair to point out, and the record clearly shows, that Dean did not follow up in any way on this. Ah, the payment that was made—Dean didn't know it, I didn't know it, nobody else knew it—apparently was being made contemporaneously that day through another source.

FROST: The next . . . the next—

NIXON: Yeah.

FROST: —the next morning, Mitchell told Haldeman that it had been paid.

NIXON: Yeah.

FROST: And in a later transcript, ah, you agree with Haldeman that he told you. You say, "Yes. You reported that to me."

NIXON: Yes. I understand.

FROST: Now, you were—

NIXON: Now, let me—

FROST: —you were very soon aware it had gone through.

NIXON: That's right. But my point is: the question we have is whether or not the payment was made as a result of a direction given by the president for that purpose. And the point is: it was not. And the point is that the next morning . . . you talk about the conversation, and, here again, ah, you probably don't have it on your notes here, but on the twenty-second, I raised the whole question of payments. And I said—and I'm compressing it all so we don't take too much of our time on this—I said, "As far as these fellows in jail are concerned, you can help them for humanitarian reasons, but you can't pay . . . but that Hunt thing goes too far. That's just damn blackmail." Now, that's in the record. And that's certainly an indication that it wasn't paid.

Nixon continued to insist he had authorized no payment, even after evidence came to light that a substantial payment to the defendants' counsel had been made on the night of March 21. It was time to confront Nixon on that transaction.

FROST: But later on that day, at some point, according to your later words to Haldeman, you were told that it had been paid.

NIXON: That . . . I, I agree that I was told that it had been paid. But what I am saying here is that the charge has been made that I directed it and that it was paid as a result of what I said at that meeting. That is . . . that charge is not true, and it's proved by the tapes, which in so many cases can be damaging, in this case they're helpful.

FROST: Well, there are two concerns to be said to that. One is: I think that my reading of the tapes tells me, trying to read in an open-minded way, that the writing, not just between the lines but on so many of the lines that I quoted, is very, very clear that you were, in fact, endorsing at least the short-term solution of paying this sum of money to buy time. That would be my reading of it. But the other point to be said is: here's Dean, talking about this hush money for Hunt, talking about blackmail and all of that. I would say that you endorsed to ratify it. But let's leave that on one side.

NIXON: I didn't endorse or ratify it.

FROST: Why didn't you stop it?

NIXON: Because at that point I had nothing to . . . no knowledge of the fact that it was going to be paid. I'd had no knowledge of the fact that, ah, the . . . what you have mentioned in the transcript of the next day, where Mitchell said he thought it'd "been taken care of"—I think that was what the words were or words to that effect, I wasn't there, I didn't . . . I don't remember what he said—that was only reported to me. The point that I make is this: it's possible it's a mistake that I didn't stop it. The point that I make is that I did consider it. I've told you that I considered it. Ah,

I considered it for reasons that I thought were very good ones. Ah, I would not consider it, ah, for, ah, the other reasons, which would have been, in my view, bad ones.

After a break, Nixon returned to the set subdued and somewhat shaken. He had apparently received advice from his own staff that the hard-line approach wasn't working and that he still had not gone far enough in conceding errors of judgment and actions that many would regard as illegal. Yet I saw no grounds to relent in my own examination of the subject, and I turned to one of the more embarrassing pieces of evidence, Nixon's tendency to coach the Watergate suspects on how to avoid perjury and other charges.

FROST: But that night, though, the night of the twenty-first. I mean, in the conversation with Colson after you'd been exchanging dialogue about getting off the reservation and so on, Colson said to you something about the fact that "It's the stuff after the cover-up. I don't care about the people involved in the cover-up; it's the stuff after that's dangerous, Dean and other things, and the things that have been done." And you said, as I'm sure you know, "You mean with regard to the defendants. Of course that was . . . that had to be done 'brackets (laughs),' " whatever that means. But I mean, so that night you were saying that had to be done. You were realizing that doing something for the defendants was a necessity.

NIXON: No, I don't interpret that that way at all. Ah, I, ah, I—

FROST: How do you recall it?

NIXON: I can't recall that . . . I can't recall that conversation, ah, and I can't vouch for the accuracy of the transcription on that. Ah, but I do say—

FROST: That's absolutely . . . it's an exhibit of the Watergate trial.

NIXON: —ah, that . . . the statements . . . the tapes that have been made public, on the twenty-second, with regard to my . . . and the one on the twenty-first as well, ah, with regard to the whole payments problem, ah, I think are very clear with regard to my attitude.

FROST: But on the short-term point that was an exhibit and part of the basic file at the trial, was that conversation. Colson saying, "It's the stuff after that's dangerous." And you saying, "You mean with regard to the defendants. Of course that was . . . that had to be done 'brackets (laughs).' " I mean, that's absolutely on the record and authenticated and played publicly.

NIXON: Well, I can't interpret it at this time.

FROST: One of the other things that people find very difficult to take is in the Oval Office on March the twenty-first, the coaching that you gave Dean and Haldeman on how to deal with the grand jury without getting caught and saying that "Perjury's a tough rap to prove," as you'd said earlier, "Just be damned sure you say, 'I don't remember. I can't recall.' " Is that the sort of conversation that ought to have been going on in the Oval Office, do you think?

NIXON: I think that kind of advice is proper advice for one who, as I was at that time, beginning to put myself in the position of an attorney for the defense, ah, something that I wish I hadn't had the re— felt I had the responsibility to do. Ah, but I would like the opportunity, when the question arises, to tell you why I felt as deeply as I did on that point. Ah, every lawyer, when he talks to a witness who's going before a grand jury, says, "Be sure that you don't volunteer anything. Be sure if you have any questions about anything, say that you don't recollect. Be sure that everything . . . that you state only the facts that you're absolutely sure of." Ah, now, on the other hand, I didn't tell them to say "Don't forget, if you do remember." That, then, would be suborning perjury. And I did not say that.

During his desperate battle to save his presidency, Nixon claimed that he himself had sought a full accounting of Watergate from his counsel John Dean but had been frustrated by Dean's inability or unwillingness to perform. Others too, such as Haldeman and Ehrlichman, were involved in planning this exercise which, not surprisingly, came to naught. Dean took his records to Camp David to write his report but promptly figured out that what the White House sought was a whitewash for its key personnel, with Dean going down with the ship. Small wonder he returned with only blank pages. Would Nixon now try to sell his requested investigation of evidence of his trying to get to the bottom of the scandal? Or would he concede that the internal investigation was as much of a sham as the CIA ruse and the "humanitarian" hush money business?

FROST: One of the things you repeated many times, but I suppose most memorably, or most clearly, on August the fifteenth, 1973, you said, "If anyone at the White House or high up in my campaign had been involved in wrongdoing of any kind, I wanted the White House to take the lead in making that known. On March the twenty-first, I instructed Dean to write a complete report of all that he knew on the entire Watergate matter." Now, when one looks through the record of what had gone on just before and after March the twenty-first, on March the seventeenth, the written statement from Dean, you asked for a "self-serving goddamned statement denying culpability of principal figures." When he told you that the original Liddy plan had involved bugging, you told him to omit that fact in his document and state it was for . . . the plan was for totally legal intelligence operation. On March twentieth, as I'm sure you know, you said, "You want a complete statement, but make it very incomplete." On March the twenty-first, after his revelations to you, you say, "Understand, I don't want to get all that goddamned specific." And Ehrlichman and you, when you're talking on the twenty-second and he's talking of the Dean report, he says, "And the report says, 'Nobody was involved.' " And there's several other quotes to that effect. Was that the Dean report that you described? It wasn't the same as what you described on August the fifteenth, was it?

NIXON: Well, what you're leaving out, of course, which is in the same tape that you've just quoted from, is a very, very significant statement. I said that "John Dean should make a report." And I said, we've . . . er . . . "We have to have a statement." And then I went on to say, "And if it opens doors, let it open doors." Now, with regard to the report being complete but incomplete, ah, what I meant was this—very simple—ah, I meant that he should state what he was sure of, what he knew. Because one day he would say one thing, another day he'd say something else. I didn't want him to answer. And you'll find that also in one of the tapes, I said, "Don't go into every charge that has been made, ah, go into only what you know, and particularly, go in hard on the fact," which he had consistently repeated over and over again, "No one in the White House is involved." That's what I wanted him to do.

FROST: But then you have a discussion in the meeting with Haldeman, Ehrlichman, Mitchell, Dean, um, where you're deciding what the policy's going to be. "Is it going to be a hang-out, i.e., is it going to be the whole of the truth?" And in the end, it's decided that it's going to be one of the great phrases of Watergate: "A modified, limited hang-out." Which is why I suggest the other quotes that I've quoted to you are decisive. And then Ehrlichman goes on to say, "I'm looking at the future." And he says, "Now, we already know it's a modified, limited hang-out." And you can't have a modified, limited version of the truth. I mean, it's obviously not going to be the whole of the truth. "I am looking at the future; assuming some corner of this thing comes unstuck at some time, you"—that's you—"are in a position to say, 'Look, that document I've published is a document I relied on, that is the report I relied on.' " And, you respond, "That's right." Now you've decided the document's going to be modified; it's going to be limited and then you're going to rely on that document and so you're going to be able to blame it on Dean. And it seems to me that that is consistent with all the quotes that I have quoted and not the "open door quote" that you have quoted.

NIXON: That's your opinion, and I have my opinion. Dean was sent to write a report. He worked on it, ah, and, ah, he certainly would have remembered, ah, ah, a phrase that was, let me say, ah, a lot more easy to understand than "modified hang-out" or whatever Ehrlichman said. Ah, he would have remembered, "If it opens doors, it opens doors." I meant by that I was prepared to hear the worst as well as the good.

FROST: What I don't understand about March the twenty-first is that I still don't know why you didn't pick up the phone and tell the cops. I still don't know when you found out about the things that Haldeman and Ehrlichman had done, that there was no evidence anywhere of a rebuke, but only of scenarios and excuses, et cetera. Nowhere do you say, "We must get this information direct to—" whoever it is, the head of the Justice Department, criminal investigation, or whatever. And nowhere do you say to Haldeman and Ehrlichman, "This is disgraceful conduct"—and Haldeman admits a lot of it the next day, so you're not relying on Dean—"you're fired."

NIXON: Well, could I take my time now to, to, ah, to address that question?

FROST: Mm.

NIXON: Ah, I think it will be very, ah, useful to you to know what I, what I was going through. Ah, it wasn't a very easy time. Ah, ah, I think, ah, my daughter once said that, ah, "There really wasn't a happy time in the White House, except in a personal sense, after April 30, when Haldeman and Ehrlichman left." You know, it's rather difficult to tell you, four years later, how you felt. But I think you'd like to know. Something new.

You see, I had been through a very difficult period when President Eisenhower had the Adams problem, and I'll never forget the agony he went through. Here was Adams; a man that had gone through the heart attack with him; a man that had gone through the stroke with him; a man that had gone through the ileitis with him; a man who had been totally selfless . . . but he was caught up in a web; ah, guilty? I don't know. I considered Adams then to be an honest man in his heart; ah, he did have some misjudgment, but, in any event, ah, finally Eisenhower decided, after months of indecision on it—and he stood up for him in press conferences over and over again, and Haggarty did—he decided he had to go. You know who did it? I did it.

Eisenhower called me in and asked me to talk to Sherm. And so here was the situation I was faced with: Who's going to talk to these men? What can we do about it? Well, first let me say that I didn't have anybody that could talk to them but me. I couldn't have Agnew talk to them because they didn't get along well with him. Bill Rogers wasn't happy with them either, and so, not having a vice president or anybody else and Haldeman, my chief of staff, himself being one involved, the only man that could talk to them was me.

Now, when I did talk to them, it was one of the most, ah, I would say difficult periods, heartrending—hard to use the adjectives that are adequate—experiences of my life. I'll never forget when I heard that, on April fifteenth from Henry Petersen, that they ought to resign and Kleindienst thought they ought to resign, and it took me two weeks . . . I frankly agreed, incidentally, in my own mind, that they had to go on the basis of the evidence that had been presented. Ah, but I didn't tell them that at that point. When I say, "I agreed with it." I didn't fully reach that conclusion because I still wanted to give them a chance to survive. I didn't want to have them sacked as Eisenhower sacked Adams and then have . . . and Adams goes off to New Hampshire and runs a ski lodge and is never prosecuted for anything; sacked because of misjudgment, yes, ah, mistakes, yes, ah, but, ah, an illegal act, ah, with an immoral, illegal motive? No. That's what I feel about Adams, and that's the way I felt about these men at the time.

Now, let me tell you what happened. I remember Henry Petersen coming in on that Sunday afternoon. He came in off his boat. He apologized for being in his sneakers and a pair of blue jeans and so forth, but it was very important to give me the update on what had . . . the developments that had occurred up to April 15. And he said . . . he gave me a piece of paper indicating that they had knowledge of Haldeman's participation and the $350,000, and they had knowledge of Ehrlichman's participation in ordering or . . . they indicated that Ehrlichman had told Hunt to deep . . . ah, the, ah, Gray to deep-six—

FROST: Six.

NIXON: —some papers and so forth and so on. And he said, "Mr. President, these men have got to resign. You've got to fire 'em." And I said to him, I said, "But, Henry, I can't fire men simply on the basis of charges, ah, they've gotta have their day in court. Ah, they've gotta have a chance to prove their innocence. I've gotta see more than this because they claim that they're not guilty." And Henry Petersen, very uncharacteristically—because he's a very respectful, a Democrat, career civil service, splendid man—sat back in his chair, and he said, "You know, Mr. President, what you've just said—that you can't fire a man simply on the basis of charges that have been made and the fact that they . . . their continued service will be embarrassing to you, you've gotta have proof before you do that." He said, "That speaks very well for you as a man. It doesn't speak well for you as a president." And, in retrospect, I guess he was right. So, it took me two weeks to work it out, tortuous long sessions. You've got hours and hours of talks with them, which they resisted. We don't need to go through all that agony. And I remember the day at Camp David when they came up. Haldeman came in first; he's standing as he usually does, not a Germanic Nazi storm trooper but just a decent, respected crew-cut guy. That's the way Haldeman was, a splendid man. And, ah, he says, "I disagree with your decision totally." He said, "I think it's going to eventually . . . you're going to live to regret it. But I will." Ehrlichman then came in. I knew that Ehrlichman was bitter because he felt very strongly he shouldn't resign. Although he'd even indicated that Haldeman should go and maybe he should stay. And I took Ehrlichman out on the porch at Aspen, you've never been to Aspen, I suppose? That's the presidential cabin at Camp David, and it was springtime. The tulips had just come out. I'll never forget, we looked out across . . . it was one of those gorgeous days when, you know, no clouds were on the mountain. And I was pretty emotionally wrought up and I remember that I could just hardly bring myself to tell Ehrlichman that he had to go because I knew that he was going to resist it. I said, "You know, John, when I went to bed last night," I said, "I hoped, I almost prayed I wouldn't wake up this morning." Well, it was an emotional moment, I think there were tears in our eyes, both of us. He said, "Don't say that." We went back in. They agreed to leave as it was late, but I did it. I cut off one arm and then cut off the other arm. Now, I can be faulted, I recognize it. Maybe I defended them too long; maybe I tried to help them too much. But I was concerned about them. I was concerned about their families. I felt that they, in their hearts, felt they were not guilty. I felt they ought to have a chance at least to prove that they were not guilty, and I didn't want to be in the position of just sawing them off in that way. And, I suppose you could sum it all up the way one of your British prime ministers summed it up, Gladstone, when he said that "The first requirement for a prime minister is to be a good butcher." Well, I think the great story, as far as a summary of Watergate is concerned, I did some of the big things rather well. I screwed up terribly in what was a little thing and became a big thing, but I will have to admit, I wasn't a good butcher.

FROST: Would you go further than "mistakes"? That, you've explained how you got caught up in this thing . . . you've explained your motives. I don't want to quibble about any of that, but just coming to the sheer substance, would you go further than "mistakes"? The word that seems not enough for people to understand.

NIXON: Well, what would you express?

This is the heart-stopping moment to which I referred earlier.

FROST: My goodness, that's a . . . I think that there are three things, since you asked me, I would like to hear you say, I think the American people would like to hear you say. One is "There was probably more than mistakes, there was wrongdoing." Whether it was a crime or not? Yes, it may have been a crime too. Secondly, "I did . . ." and I'm saying this without questioning the motives, right, "I did abuse the power I had as president, or not fulfil the totality of the oath of office." That's the second thing. And thirdly, "I put the American people through two years of needless agony, and I apologize for that." And I say that you've explained your motives. I think those are the categories. And I know how difficult it is for anyone, and most of all you, but I think that people need to hear it, and I think, unless you say it, you're going to be haunted for the rest of your life.

NIXON: I well remember that when I let Haldeman and Ehrlichman know that they were to resign, that I had Ray Price bring in the final draft of the speech that I was to make the next night. And I said to him, "Ray," I said, "if you think I oughta resign," I said, "put that in too because I feel responsible." Even though I did not feel that I had engaged in these activities consciously. Ah, insofar as the knowledge of or participation in the break-in; the approval of hush money; the approval of, ah, clemency, et cetera. The various charges that have been made. Well, he didn't put it in. And I must say that, at the time, I seriously considered whether I shouldn't resign. But on the other hand, I feel that I owe it to history to point out that, from that time on April 30 until I resigned on August 9, I did some things that were good for this country. We had the second and third summits. I think one of the major reasons I stayed in office was my concern about keeping the China initiative, the Soviet initiative, the Vietnam fragile peace agreement, and the added dividend, the first breakthrough in moving toward not love but at least not war in the Middle East.

FROST: You've—

NIXON: And now, coming back to the whole point of, ah, whether I should have resigned then and how I feel now. Let me say, I . . . I just didn't make mistakes in this period. I think some of my mistakes that I regret most deeply came with the statements that I made afterwards. Ah, some of those statements ah, were misleading, ah, I notice, for example, the editor to *The Washington Post,* the managing editor, Ben Bradlee, wrote a couple, three months ago, something to the effect that, as far as his newspaper was concerned, he said, "We don't print the truth. We print what we know. We print what people tell us and this means that we print lies."

Ah, I would say that the statements that I made afterwards were, on the big issues, true. That I was not involved in the matters that I have spoken to you about; not involved in the break-in; that I did not engage in the . . . and participate in, or approve, the payment of money or the authorization of clemency, which of course were the essential elements of the cover-up. That was true. Ah, but the statements were misleading in exaggerating that enormous political attack I was under. It was a five-front war with a fifth column, ah, ah, I had a partisan Senate committee staff. We had a partisan, ah, special prosecutor staff; we had a partisan media; we had a partisan Judiciary Committee staff in the fifth column. Now, under all these circumstances, my reactions in some of the statements and press conferences and so forth after that, I want to say right here and now, I said things that were not true. Most of them were fundamentally true on the big issues but without going as far as I should have gone and saying perhaps that I had considered other things but had not done them.

FROST: Well . . . you mean that—

NIXON: And for all those things I have a very deep regret.

FROST: You got caught up in something—

NIXON: Yeah.

FROST: —and then it snowballed—

I could feel myself draining emotionally, and I could see that Nixon was grappling with his own deepest feelings.

NIXON: It snowballed. And it was my fault. I'm not blaming anybody else. I'm simply saying to you that, as far as I'm concerned, I not only regret it—I indicated my own beliefs in this matter when I resigned. People didn't think it was enough to admit mistakes, fine. If they want me to get down and grovel on the floor, no. Never. Ah, because I don't believe I should. On the other hand, there are some friends who say, "Just face them down. There is a conspiracy to get you." There may have been. I don't know what the CIA had to do. Some of their shenanigans have yet to be told, according to a book I read recently. Ah, I don't know what was going on in some Republican, some Democratic circles, as far as the so-called impeachment lobby was concerned. However, I don't go with the idea that there . . . that what brought me down was a coup; a conspiracy, et cetera, et cetera, et cetera. Ah, I brought myself down. I gave them a sword. And they stuck it in, and they twisted it with relish. And I guess if I'd been in their position, I'd have done the same thing.

FROST: But what I'm really saying is that in addition to the untrue statements that you've mentioned, could you just say, with conviction I mean, not because I want you to say it, that you did do some covering up? We're not talking legalistically now, I just want the facts. I mean, that you did do some covering up? That there were a series of times when, maybe overwhelmed by your loyalties or whatever else, but as you look back at the record, you behaved partially protecting your friends, or maybe yourself, and that, in fact, you were, to put it at its most simple, a part of a cover-up at times?

NIXON: No, I . . . I again, I again respectfully will not quibble with you about the use of the terms. However, before using the term, I think it's very important for me to make clear what I did not and what I did do. And then I will answer your question quite directly. Ah, I did not, ah, in the first place, ah, commit a . . . the crime of obstruction of justice. Because I did not have the motive required for the commission of that crime.

FROST: We've had our discussion on that, and we disagree on that, but that's—

NIXON: The lawyers can argue that. I did not commit, in my view, an impeachable offense. Now, the House has ruled overwhelmingly that I did, ah, of course, that was only an indictment and would have to be tried in the Senate—I might have won, I might have lost—but even if I'd won in the Senate by a vote or two, I would have been crippled and the . . . in any event, for six months the country couldn't afford having the president in the dock in the United States Senate, and there can never be an impeachment in the future in this country without voluntarily impeaching himself. I have impeached myself. That speaks for itself.

FROST: How do you mean, "I have impeached myself"?

NIXON: By resigning. That was a voluntary impeachment. And, ah, now what does that mean in terms of whether I, ah . . . you're wanting me to say that I am . . . participated in an illegal cover-up? No. Now, when you come to the period—and this is the critical period—that when you come to the period of March 21 on, when Dean gave his legal opinion, ah, that certain things, actions taken by Haldeman, Ehrlichman, Mitchell, et cetera, and even by himself, amounted to a legal cover-up and so forth, then I was in a very different position, and, during that period, I will admit that I started acting as lawyer for their defense. I will admit that acting as lawyer for their defense, I was not prosecuting the case. I will admit that during that period, rather than acting primarily in my role as the chief law enforcement officer in the United States of America, or at least with responsibility for the law enforcement, because the attorney general is the chief law enforcement officer, but as the one with the chief responsibility for seeing that the laws of the United States are enforced, that I did not meet that responsibility.

I could not imagine Nixon going any further than he had to this moment. But I was wrong.

NIXON: And to the extent that I did not meet that responsibility, to the extent that, within the law and in some cases going right to the edge of the law in trying to advise Ehrlichman and Haldeman and all the rest as to how best to present their cases because I thought they were legally innocent, that I came to the edge, and, under the circumstances, I would have to say that a reasonable person could call that a cover-up. I didn't think of it as a cover-up. I didn't intend it to cover up. Let me say, if I'd intended to cover up, believe me, I'd have done it. You know how I could have done it so easily? I could have done it immediately after the election simply by giving clemency to everybody and the whole thing would have gone away. I couldn't do that because I said, "Clemency is wrong."

But now we come down to the key point. And let me answer it in my own way about "How do I feel about the American people?" I mean, ah, how, ah, whether I should have resigned earlier or what I should say to them now. Well, that forces me to rationalize now and give you a carefully prepared and cropped statement. I didn't expect this question, frankly, though, so I'm not going to give you that, but I can tell you this—

FROST: Nor did I.

NIXON: —I can tell you this: I think I said it all in one of those moments that you're not thinking. Sometimes you say the things that are really in your heart. When you're thinking in advance, then you say things you know are tailored to the audience. I had a lot of difficult meetings those last days before I resigned, and the most difficult one, and the only one where I broke into tears, frankly, except for that very brief session with Ehrlichman up at Camp David, it was the first time I had cried since Eisenhower died. I met with all of my key supporters just a half hour before going on television. For twenty-five minutes, we all sat around in the Oval Office; men that I'd come to Congress with; Democrats and Republicans, about half and half, wonderful men. And, at the very end, after saying, "Well, thank you for all your support during these tough years. Thank you for the, particularly for what you've done to help us end the draft and bring home the POWs and have a chance for building a generation of peace"—which I could see the dream that I had possibly being shattered—"and thank you for your friendship, little acts of friendship over the years. . . ." You know, you sort of remember, you know, with a birthday card and the rest. Then, suddenly, you haven't got much more to say, and half the people around the table were crying. Les Aarons, Illinois, bless him, he was shaking, sobbing, and, ah, I get . . . just can't stand seeing somebody else cry, and that ended it for me. And I just, well, I must say I sort of cracked up, started to cry, pushed my chair back, and then I blurted it out, I said, "I'm sorry. I just hope I haven't let you down."

Well, when I said, "I just hope I haven't let you down," that said it all.

I had.

I let down my friends.

I let down the country.

I let down our system of government and the dreams of all those young people that ought to get into government but think it's all too corrupt and the rest.

Most of all, I let down an opportunity that I would have had for two and a half more years to proceed to great projects and programs for building a lasting peace, which has been my dream, as you know from our first interview in 1968, before I had any thought I might even win that year. (I didn't tell you I didn't think I might win, but I wasn't sure.)

Yep, I . . . I, I let the American people down, and I have to carry that burden with me for the rest of my life.

My political life is over.

I will never yet, and never again, have an opportunity to serve in any official position. Maybe I can give a little advice from time to time.

And so, I can only say that, in answer to your question, that while technically I did not commit a crime, an impeachable offense—these are legalisms.

As far as the handling of this matter is concerned, it was so botched up.

I made so many bad judgments. The worst ones, mistakes of the heart, rather than the head, as I pointed out.

But let me say, a man in that top judge . . . top job, he's gotta have a heart.

But his head must always rule his heart.

It was at this point that our broadcast interview ended. Nixon had traveled a long and circuitous route from denial and defiance to acceptance and admission. As both his staff and mine predicted, the sight of a contrite, nonthreatening Nixon was balm to the political community, which gradually over the ensuing years accepted the return of its prodigal son to membership in goodish standing. But Richard Nixon himself was not expecting anything of the sort . . .

FROST: I think you've said it all, really. The, you're saying, if I understand it right, Mr. President, you said, "It's a burden that you've got to carry with you for the rest of your life." I think it may be a little, a little lighter after what you've said and—

NIXON: I doubt it. I remember very well, ah . . . the night before I resigned, Eddie Cox, a wonderful young man, Tricia's husband, you know, sort of Princeton and on the tennis team and fine family in New York and all the rest, good lawyer, one of the great law firms in New York, God, I wish we'd had him in the White House, had he been old enough at the time. And I said, "Well, at least"—and I knew this wasn't true, but I was just giving him as the devil's advocate, the proposition—I said, "Well, at least, Ed," he knew I'd made the decision, "this cuts it off. We'll go out to California and they'll leave us alone." I've never seen him . . . he's a very well-contained boy, he said, "Look, oh, no, they won't." He said, "You don't know these people." He says, "I know them." What he was speaking of was the staff of the special prosecutor, Mr. Jaworski's staff. As a matter of fact, Mr. Jaworski told Al Haig on several occasions that he inherited from Archibald Cox a lot of hot rods who were pushing for things that he thought went too far and that he just couldn't control 'em.

But, be that as it may, Ed said, "Let me tell you something about 'em. I worked in the U.S. attorney's"—and, incidentally, he had also worked for Ralph Nader, so he's had a pretty good experience—but "I worked in the U.S. attorney's district . . . office in New York. I went to school, with these . . . some of these people at the Harvard Law School, and I know something about 'em. They're tough; they're smart. But, most of all, they hate you with a passion. Mostly because of the war, and some because of other reasons. And, they and others like them and the press, they're going to hound you, they're going to harass you for the rest of your life." And as we conclude this, I can say they have and they will, and I will take it, I hope, like a man.

8

THE HUSTON PLAN

In questioning Mr. Nixon about the so-called Huston plan, the purpose of which was to counter domestic political violence in the United States, we were dealing with an issue that had furnished grounds for a second article of impeachment against Mr. Nixon. Yet there were stark historical differences with Watergate that had to be reflected in my questions. Many administrations had challenged constitutional liberties during periods of crisis: Lincoln suspended habeas corpus in certain areas, Wilson and his immediate successors prosecuted war dissenters who arguably posed no security threat to the nation. John Adams passed the Alien and Sedition Acts. Further, such actions were justified by the chief executive involved as consistent with the implied powers of the presidency. Some presidential actions were upheld, others promptly rescinded or struck down, but the issues always involved interpreting the Constitution. Contrast this to Watergate, which began as a "third-rate burglary" and soon involved hush money payments, subornation of perjury, and the whole plethora of violations known as "the cover-up." My questioning, therefore, had to reflect these historical assertions of presidents while not giving an inch on substance. Watergate, by contrast, would later be more nearly akin to the cross-examination of a defendant—at least on the first day.

FROST: You called a meeting on June the fifth, 1970, about the Huston plan and eventually approved it in July. It got your okay on July the fourteenth, didn't it? And in the Huston plan it stated very clearly, with reference to the entry that was being proposed, it said very clearly, use of this technique is clearly illegal, it amounts to burglary . . . however, it is also one of the most fruitful tools and it can produce the type of intelligence which cannot be obtained in any other fashion. Why did you approve a plan that included an element like that . . . that was clearly illegal?

NIXON: Because as president of the United States . . . ah . . . I had to make a decision, as has faced most presidents, in fact, all of them, ah . . . in which, ah . . . the national security in terms of a threat from abroad, ah . . . and the security of the individual . . . individual violence at home had to be put first. Ah . . . I think Abraham Lincoln has stated it better than anybody else, as he does in so many cases. When he said, "Must a government be too strong for the liberties of its people? Or too weak to defend or maintain its own existence?" That's the dilemma that presidents have had to face, ah . . . Roosevelt had to face it in World War II. Truman and Eisenhower in the Cold War period. Kennedy and Johnson as Vietnam began to come in. And Kennedy, of course, even before Vietnam began to escalate, had the beginning of the violent racial disturbances . . . ah . . . which led to some activities in this category. Now let's first, let's second understand what the surreptitious entry is limited to. You will note that a surreptitious entry in cases involving national security and specifically mentions, ah . . . two, ah . . . groups of, ah . . . internal organizations who had no foreign connections as far as we know. Ah . . . the Weathermen and the Black Panthers.

Now, why were we concerned? Let's look at the year, 1970. We had a situation where thirty-five thousand people, ah . . . had been victims of assaults. A number of them had been killed. It was a year in which we had had, ah . . . sixteen airplane hijackings. There had been about eleven the year before. Ah . . . but most significantly, it was a year in which there had been thirty thousand bombings and fifty thousand . . . I mean, sorry, three thousand bombings, three thousand bombings and fifty thousand bomb threats . . . which caused, ah . . . the evacuation of buildings. Ah . . . it was a year of turbulence in American society. Ah . . . '68 . . . '69 . . . '70 . . . the residue of the terrible period of '68. Washing over into '69 and continued through '70 and then, thank God, began to go down in '71 and '72, when calm was restored to the campuses. The cities did cease to be burned, and bombings did go down. And while we've argued about our crime statistics, where at least in '72 there was a decrease rather than an increase. All right, now, now in 1970, in the middle of 1970, ah . . . we were faced with a situation here, first, where the intelligence agencies weren't working together. Ah . . . there were CIA . . . was not speaking to the FBI . . . the NSA, the National Security Agency, which of course does all of our cryptic, graphic work. That's the highly sensitive, technical work, you know, to break codes and that sort of thing . . . had very little communication with the other two. Ah . . . under the circumstances I felt that we had to coordinate these activities and get a more effective program for dealing with, first, foreign-directed, ah . . . espionage, ah . . . or foreign-supported, ah . . . subversion. And in addition with domestic groups that used and advocated violence.

FROST: But was there ever such a thing as really foreign support in subversion in the sense that—I mean, Johnson had investigated and the CIA investigated it and so on, and all those reports have always said that search as they may and wish as they may to find, ah . . . foreign money, Cuban money, Russian money in some of the dissenting groups in this country, they've never really found it, have they—

NIXON: As a matter of fact, you are absolutely correct. Because I specifically asked that that be investigated. I wanted to know. Ah . . . I found that what had happened, ah . . . was members of these groups were used by the Cubans, ah . . . by the North Koreans and, ah . . . sometimes by others and by the North Vietnamese . . . they made trips there and so forth and so on. As far as ever getting proof of any significant amounts of foreign money to those specific groups, the Weathermen and the Black Panthers . . . we found very little evidence to back it up. Ah . . . now, as far as the Communist Party in the U.S.A. was concerned, ah . . . no question about direct subsidy and a substantial one from the Soviet Union. Not from the government of the Soviet Union but from the Communist Party of the Soviet Union. That's been going on for years . . . that's been common practice. Now, insofar as this . . . these groups are concerned, though, let me say that surreptitious entry had two purposes. One, ah . . . for example, take the code business, I . . . I'll never forget the . . . the admiral . . . Admiral Gaylor, as a matter of fact, he happened to have the honor of being the admiral who welcomed the POWs back when they landed at Clark Field back that memorable day on February the eleventh, '73 . . . but anyway, Admiral Gaylor sat there and was talking about this whole matter of surreptitious entry. He said, "You know, just one entry, ah . . . can obtain information in a day that it might take us months to obtain if we don't have the entry." So there's no question about the need for that kind of entry. Entry into a foreign embassy, ah . . . for the purpose of code breaking and arrest . . . and that's mentioned, if you note . . .

If you read the Huston report carefully, which of course I know you have, ah . . . but now, as far as the Weathermen and the others were concerned, there the problem was one that I have just mentioned, ah . . . they were engaged in violence. They practiced . . . they not only preached it, but they practiced it. They bomb and everything else. I think, for example, the fact . . . you mentioned . . . and it's just triggered my memory just now . . . that on July the fourteenth, when I approved the plan, ah, then it was a week later when I disapproved it when Hoover, I guess, came in or two weeks later, Hoover saw Mitchell, and Mitchell passed on his objections to the plan, and we decided not to put it into effect, and a month later, you know what happened? Well, they blew up, when I say that "they," one of these groups blew up a building at the University of Wisconsin, Madison . . . and a twenty-five-year-old student was killed and two others were injured . . . and the property damage was of course in millions. If we'd had the plan into effect, maybe he'd be living.

That's why when, ah . . . people sometimes raise questions about all of this kind of activity that went on, ah . . . in 1963, when this whole area of surreptitious entry and so forth reached its peak in 1963, 1964, let me say that if it had succeeded in getting one lead it could have uncovered one of those involved in the Kennedy assassination, assuming that Oswald had any coconspirators. Or even uncovering him. It sure would have been worth it. That's what a president is confronted with, and so as far as I was concerned, I considered that the president had not only the power but he had the responsibility, in this instance, to put the safety of citizens, ah . . . above, ah . . . the legal technicality that was involved. But on a very limited basis.

FROST: But why did Hoover oppose the plan?

I barely suppressed a smile as I asked that question. For here was the incongruity of the liberals' nemesis J. Edgar Hoover refusing to implement a presidential plan that would have had his agents cavorting through the rooms and makeshift offices of domestic groups like the Weathermen and Black Panthers. Was Hoover a born-again civil libertarian? Or, having sustained himself through seven presidential administrations, was he just a damned good pol?

NIXON: Hoover is . . . had perhaps the best public relations sense of anybody in Washington. For all the years that he was there . . . it was true in the eight years I knew him when Eisenhower was president and the years when I was president until he died in 1972. He had become—and particularly as he got older—he began reading his press clippings. Well . . . he always read them, and usually they were favorable. But he began reading them more intently, and they began to be unfavorable. And he was very sensitive about that because he wanted the favorable press. They became unfavorable because basically, ah . . . in the fifties, it was popular to be against Communist subversion. It was popular to be against spies. It was popular to be against espionage. It was popular to be against those that stole papers and then needless to say have them printed in the papers.

Today, when you have a situation where those that steal documents are made heroes and papers that print them, ah . . . get Pulitzer Prizes for them . . . me, like Hoover or others who are on the other side, don't look that good. So he was extremely sensitive, ah . . . starting in about 1966 about any activity that would be undertaken. As a matter of fact, in this whole area, Hoover's attitude was "the Bureau doesn't want to do this," but he would have no objection if other agencies were doing it. So he didn't mind having the CIA do it. But obviously the CIA couldn't do it because they don't have authority to or are not supposed to have authority to do things in the United States.

FROST: —jobs or anything else—

NIXON: Exactly . . . that's right . . . in the United States . . . Now, I can't say that they did or didn't because I have found since that the CIA and the FBI, too, have done some things that I didn't know and I assume other presidents may not have known of.

FROST: But that particular point, Hoover's argument, was what, then?

NIXON: Public relations . . . that it would be discovered. He felt that it might be usable. But he thought that the possibility of its being public and the heat that we would get publicly would outweigh the usefulness.

FROST: But there's an interesting schizophrenia in that case. You said the reason you instituted the plan was, ah . . . that you put safety of citizens above everything else. And the reason Hoover objected to the plan was public relations, so in the end you put public relations above safety of citizens.

NIXON: When . . . I did—

FROST: Yeah—

NIXON: When I took Hoover's objections?

FROST: Yeah.

NIXON: I did, yes. Now, I did it for another reason, however. It wasn't the public relations . . . it wasn't Mr. Hoover's argument. No, I missed your question. I want to get it right here. Hoover's argument was public relations. But the point was that, without Hoover's cooperation, the whole house of cards fell. There couldn't be a Huston plan, we couldn't implement it because the CIA or the NSA couldn't do the surreptitious entries. They had to be done by the FBI, and if J. Edgar Hoover wouldn't go along, there was no way we could implement it. So that was it. I think that Hoover, in this case, gave bad advice. I wish, as a matter of fact, in retrospect, that we had moved forward with the plan, that Hoover would have been willing. Move forward with it on the very restricted basis that it has been described here. That's far better than to have had it go forward, some of these things go forward, as they had in previous administrations and, ah, apparently as they did even after this without any plan, without any control. I think the president ought to know when this kind of activity takes place. And that he therefore can exercise control over it rather than having it be in a situation when he doesn't know that the CIA, for example, opened mail.

FROST: Given that you've said that Hoover had a virtual veto. By virtue of his position on this . . . didn't you feel at this moment while you were telling the networks and the media who was president . . . didn't you feel Hoover needed to know who was president?

NIXON: Let's be candid with regard to the relations that the president has with his cabinet and with the top appointees. Scrape away some of the double-talk that ah . . . when you're in office you have to use, frankly, in order to keep the team from tearing itself apart . . . and when you're out of office you don't have to use. But basically the situation was that, ah . . . Edgar Hoover still, at that time, despite the fact of his age, despite the fact that morale in the Bureau was not as good as it had been previously, Edgar Hoover, in my view, was still the best qualified man to be head of the Bureau. I felt that he was, and second, ah . . . I also knew that in terms of support in the Congress and in the country his support was still overwhelming.

And as president, I was not about to take him on unless it had been on an issue so big, so overwhelming, like the issue that MacArthur and Truman had when Truman had to let MacArthur go because they were determining who's going to be commander in chief and a decision in a war. This decision was one where I really didn't agree with Hoover but, on the other hand, where I didn't feel that the gain of overruling him and frankly causing him to resign, ah . . . which he might have done was worth, ah . . . the results which we could have gotten by implementing the plan.

FROST: So it was too big a risk to take.

NIXON: The plan . . . the plan was tough . . . too . . . in many respects . . . how you gonna stop bombers? Ah . . . how you going to, ah . . . break codes? How you gonna stop espionage . . . leaks, etc., etc. How you gonna deal with subversives? These are all very difficult matters, and no plan, Huston plan, which provided for what you call surreptitious entry which means going in—

FROST: —burglaries—

NIXON: Yes, and, ah . . . burglary is the word, and another way to put it is that if you break into an individual's office and you take out the papers or you photograph the papers and you get the information that you want . . . need . . . in order to convict that person or to get information that may avoid some sort of illegal action in the future.

Now for one of the classic moments in the entire series. We could have continued a philosophical discussion of presidential power. But it seemed to me we were dealing with a simpler concept, the power of the president who waives constitutional protection where he deems the national interest at stake. Not the easiest topic to make come alive on television, so I came up with a question that couldn't be simpler, hoping that Nixon would take the bait and provide a truly memorable response.

FROST: Yeah . . . but I mean, that action, as Huston warned, is illegal. Now, when you were concerned about street crime and so on, you went to Congress and got laws passed and so on. Wouldn't it have been better here . . . though to have done what you were going to do legally rather than doing something that was illegal and seizing evidence in that way and all of that. In retrospect, wouldn't it have been better to combat that crime legally rather than adding another crime to the list?

NIXON: Basically the, ah . . . proposition you've just stated in theory is perfect . . . but in practice just won't work. Ah . . . let's talk about the CIA for a moment. I remember when Allen Dulles died . . . I was an old friend of his. I'd worked with him for years. I made a statement about him . . . and I said the great tragedy of a man who has served in the CIA is that his successes will never be known. His failures will always be advertised. Because the Bay of Pigs for example, a covert operation, had it been successful, I mean, it wouldn't have been known until years later . . . if at all . . . but if it failed . . . it becomes known.

Now, in this case . . . to get legislation, specific legislation, to have warrantless entries for the purpose of obtaining information and the rest would not only have raised an outcry but it would have made it terribly difficult to move in on these organizations because basically, ah . . . they would be put on notice by the very fact that the legislation was on the books that they'd be potential targets. And actions are either going to be covert or not. That's why I think that the Pike Committee and the Church Committee castrating the CIA has rendered a terrible disservice to the country, and I am incidentally very glad to see President Carter has made some strong statements, as has the new head of the CIA, to the effect that, that there are some actions that have to be covert, ah . . . by covert . . . let me put it this way—

FROST: Or in this case illegal.

NIXON: Let me say that it is legal in my view. If a president, or let's put it another . . . was it legal for Lincoln to deny the right of habeas corpus in civil war? Was it legal, for example, to, ah . . . shoot down the draft dodgers or not draft dodgers but those who were objecting to the draft in New York City and so forth and so on . . . and the point is maybe so . . . maybe not . . . was it legal, for example, to take tens of thousands, and I grew up with some of these young people and I know of loyal American-born Japanese, and keep them in a camp out here in California during World War II? Was that legal? Was it right? In retrospect, no.

But on the other hand, a president makes an order, has to issue an order where you are involved in war or in the case of our people the Weathermen, the Black Panthers, if you have thirty thousand bombings on your hand and thirty-five thousand policemen being injured a year . . . I mean three thousand bombings and thirty-five thousand policemen being injured a year . . . you gotta do something about it. The question is maintaining the proper balance. Ah . . . and what you do in these instances, ah . . . cannot always be public. And certainly in the intelligence field, you, ah . . . the suggestion that in the intelligence field that, ah . . . we've just got to make it all public. For example, our aid to Hussein, the new head of the CIA pointed out, did incalculable damage to have that leak. Do you know why? Because that practically puts Hussein in the American pocket. It makes him look as if he's an American puppet, even though his government has received some aid over the year.

FROST: So that there—

NIXON: It should be covert . . . the Congress should be informed on a very confidential basis and limited basis. But it should be covert.

FROST: But I mean—

NIXON: —and we've got to have more of it in the future, ah . . . if we're going to meet the covert activities of our potential adversaries in this world. There's one thing you have to have in mind . . . and I'm finished on this point . . . I'll state it very simply. We've been talking about war and peace a great deal . . . and I've indicated my own belief that because of the balance of terror in the world where nuclear weapons are concerned, provided we maintain a conventional balance as well, it is not likely that we will have war in a conventional sense of great armies marching across borders . . . that is my belief . . . on the other hand . . . where the Communists are concerned . . . as the danger of war goes down, conventionally speaking, the danger of conquest without war through subversion geometrically goes up and that because they are still as an article of faith—committed to the proposition of attempting to take over other countries . . . to impose Communist governments if they can get it.

Now, you've gotta combat that some way . . . and I would prefer to combat it all out in public on living color television, ah . . . so that everybody can see it . . . and put all of our CIA agents out so that everybody can see them and advertise all the money that we're providing, ah . . . if we'd play that kind of a game, we're gonna lose. Now, I don't mean that we play the game the way they do . . . but there is a ground in between, and we have followed that ground pretty well since World War II, and we should continue to follow it in the future.

FROST: So, what in a sense you're saying is that there are certain situations and the Huston plan or that part of it was one of them where the president can decide that it's in the best interest of the nation or something and do something illegal.

NIXON: Well, when the president does it . . . that means that it is not illegal.

I could scarcely believe my ears. Nixon had just made the statement that—apart from Watergate—became what was probably the most quoted sentence from the Nixon interviews. My task now was to keep him talking on this theme for as long as possible.

FROST: By definition—

NIXON: Exactly . . . exactly . . . if the president . . . if, for example, the president approves something . . . approves an action, ah . . . because of the national security or in this case because of a threat to internal peace and order of, ah . . . ah . . . significant magnitude . . . then . . . the president's decision in that instance is one, ah . . . that enables those who carry it out to carry it out without violating a law. Otherwise they're in an impossible position.

FROST: So that the black-bag jobs that were authorized in the Huston plan . . . if they'd gone ahead, would have been made legal by your action?

NIXON: Well . . . I think that we would . . . I think that we're splitting hairs here. Burglaries per se are illegal. Let's begin with that proposition. Second, when a burglary, as you have described a black-bag job, ah . . . when a burglary, ah . . . is one that is undertaken because of an expressed policy decided by the president, ah . . . in the interests of the national security . . . or in the interests of domestic tranquillity . . . ah . . . when those interests are very, very high . . . and when the device will be used in a very limited and cautious manner and responsible manner . . . when it is undertaken, then, then that means that what would otherwise be technically illegal does not subject whose who engage in such activity to criminal prosecution. That's the way that I would put it. Now, that isn't trying to split hairs . . . but I do not mean to suggest the president is above the law . . . what I am suggesting, however, what we have to understand, is, in wartime particularly, war abroad, and virtually revolution in certain concentrated areas at home, that a president does have under the Constitution extraordinary powers and must exert them with . . . as little as possible.

That's why, for example, I think in the Johnson administration, they went totally overboard in their huge program of military surveillance . . . You're probably familiar with that where this was the program that was developed by Mr. Califano and others for President Johnson as a result of the riots in 1967 and we discovered it when we got there and there were a billion and a half . . . a billion and a half records that over fifty-seven government agencies had collected. We knocked it off in 1970 . . . we just got the Army out of that sort of business. Ah . . . they had gone overboard, in my opinion. Ah . . . I felt that, ah . . . it was better to leave it to the professionals . . . to the FBI. That's why I would have preferred that Mr. Hoover would have undertaken some other things rather than the line that he did.

FROST: But just so we understand this. Equally, it would apply, presumably, to these burglaries that we were talking about, presumably . . . The people would not be open to criminal prosecution at the end? Equally, it would . . . in the theoretical case, where the action ordered by the president was a murder, it would also apply?

NIXON: No . . . I think . . . that the reason that when I answered the question earlier, I tried to put it in more—

FROST: Well, why would the burglaries not be liable to—

NIXON: Let us . . . let us suppose this . . . if, for example, President Roosevelt had decided that, ah . . . the assassination of Hitler before World War II would save five or six million Jews from extermination, ah . . . I'm not sure that that wouldn't be an awfully tough call—

FROST: No . . . but, ah—

NIXON: Now, my point is . . . as far as the assassination is concerned—

FROST: We're talking about dissent in this country.

NIXON: Dissent in this country?

FROST: I mean, in terms of these burglaries, you were saying—

NIXON: Oh, no . . . no . . . no.

FROST: If these burglaries were happening, they wouldn't be liable to criminal prosecution.

NIXON: Well . . . that's oh . . . I—

FROST: Well, what about if a murder was ordered in this country, would the presidential—

NIXON: No . . . absolutely not.

FROST: —shield also protect the murderer?

NIXON: No . . . no . . . because I don't know anyone who has been president, or is now, who would ever have ordered such an action.

FROST: No . . . nor do I have evidence.

NIXON: And the Huston plan . . . as you know, is very carefully worded that in terms of how limited it is to be—

FROST: But no, all I was saying was, where do we draw the line? if you're saying that a presidential fiat can in fact mean that someone who does one of these black-bag jobs, these burglaries, is not liable to criminal prosecution . . . why shouldn't the same presidential power apply to somebody who the president feels, in the national interest, should murder a dissenter? I'm not saying it's happened. I'm saying what's the dividing line between the burglar being liable to criminal prosecution and the murderer?

NIXON: Because as you know from many years of studying and covering the world of politics and political science, there are degrees, there are nuances, ah . . . which are difficult to explain but which are there. Ah . . . as far as this particular matter is concerned, ah . . . each case has to be considered on its own merits, and what I am saying here . . . considering the situation we were faced with at the time . . . it was wartime at home and, ah . . . virtually revolution in some areas of the country, ah . . . also at home . . . wartime abroad and at home and virtual revolution in some parts of the country at home. Under those circumstances, I think this was the least that the president could have done and the most that he probably could have done. As a matter of fact, that's why it was done so carefully. That's why we got in Helms, we got in Hoover, we got in Gaylor, we got in, ah . . . all of the people to get their input. Bob Finch was there—

FROST: So that you—

NIXON: So that it was a combined unanimous decision except for Hoover's dissent—

FROST: So that in other words, really, the only dividing line, really, you were saying, between the burglary and murder, again there's no subtle way to say that there was a murder of a dissenter in this country because we don't have any evidence to that effect at all . . . but the point is, just the dividing line is that, in fact, the dividing line is the president's judgment?

NIXON: Yes . . . and, ah . . . the dividing line, ah . . . just so that one does not get the impression, ah . . . that a president can run amok in this country and get away with it . . . ah . . . we have to have in mind that a president has to come up before the electorate . . . we also have to have in mind, ah . . . that a president has to get appropriations from the Congress . . . we have to have in mind, for example that, as far as the CIA's covert operations are concerned . . . as far as the FBI's covert operations are concerned, ah . . . through the years, ah . . . they have been disclosed on a very limited basis to trusted members of the Congress . . . I don't know whether it can be done today or not.

FROST: But on the other hand—

NIXON: And that's really strange—

FROST: Yes, I don't think that reading the documentation that it was even intended, was it, that the Huston plan and the black-bag robbery should be revealed to the electorate?

NIXON: No . . . no, these were not, that's correct. That's correct.

Having dealt with the question of the president's authority to order murder, I wanted to give Nixon one more shot at identifying any past president who had claimed the inherent power to commit burglaries and other apparent Fourth Amendment violations. This he could not do.

FROST: In fact, is there any single case at all of where any former president has personally approved black-bag jobs?

NIXON: I can't speak for any of them. None have ever indicated to me they have. I would assume that if they were intelligent people, they had to know that embassies had been targets for black-bag jobs because Edgar Hoover used to talk freely about . . . ah, well . . . he talked freely at least to his intimates about, ah . . . having, ah . . . information directly from an embassy. And he didn't get it by attending a cocktail party either.

FROST: But I don't think any former president has personally approved black-bag jobs for places other than embassies, certainly, has he?

NIXON: He may not have approved those, and he may not have known about them. I'm only suggesting that if he was perceptive, he probably . . . he could have known. Ah . . . I didn't approve them, of course, and I think in this instance, however, the Huston plan, ah . . . was basically a better approach. It did formalize it. Ah . . . I think requiring presidential approval was a good idea.

FROST: And that's why your presidential approval was unique, though—

NIXON: It was unique.

FROST: —for the black-bag jobs?

NIXON: You see, the point was, it had never been, and incidentally, I hadn't . . . I didn't know at the time . . . let me be . . . in fairness to my predecessors say, I didn't know at the time that, ah . . . for example that there had been mail openings going on for years. I didn't know at the time there had been black-bag jobs. I was only told that there had been . . . but that they'd been discontinued in 1966. Ah, ah . . . and, frankly, I didn't know at the time whether former presidents had known about them or had not known about them. All that I did believe in . . . I believe in an orderly process. I believe in a control of the process, and I'd much rather have so-called black-bag jobs and surreptitious entries. I'd rather have the director of the CIA or the FBI or the NSA or whatever group was involved in any of this kind of activity; the NSA overhearing conversations, for example, and the CIA doing whatever the CIA does, I'd much rather have the president know about everything and be a restraining force, rather than being kept in the dark on it.

FROST: Because it may be illegal, but at least it makes it tidier.

NIXON: Yes, and we must be under no illusions, ah . . . it may be that we can find a way, find a way to do these things. When I say "do these things," you've talked only about black-bag jobs, which of course is far more interesting for our audience, because you see that kind of thing on television and the rest. But some of this cryptographic overhearing that they do, ah . . . the NSA, I mean, they can monitor your conversations whenever you and I talk. When you're in London, incidentally, remember we're on the air.

FROST: Right.

NIXON: I don't think I could go through this room today and guarantee you or me that we weren't being bugged. It's so sophisticated. It's so advanced.

9

CHILE

The story of Chile and Richard Nixon can be told in a nutshell. In September 1970, a Marxist candidate named Salvador Allende finished first among all candidates for the Chilean presidency with 37 percent of the vote. Under Chilean tradition, the race was then decided by vote of the legislature, and when that happened, the leading candidate was invariably given the office. Upon Allende's first-place finish, President Nixon immediately instructed the CIA to see what it could do to have the legislature disavow past precedent and elect a non-Marxist president. That didn't work. Nor did subsequent attempts by the United States to organize a coup to remove Allende from office.

The arguments against disturbing the results of the Chilean campaign were many. Most important, Allende did not threaten to establish a Marxist state but rather served as an elected Marxist candidate within a functioning democracy. Second, a coup would lead to military rule and the prospect of widespread executions and a societywide denial of human rights. Third, the United States would undoubtedly be blamed for the coup, not only within Chile but throughout the hemisphere. Nixon saw things the other way, as was well articulated in our interview. While making the case for a U.S.-engineered coup d'état, he denied that the 1970 coup, in which Allende lost his life, had been made in the U.S.A., saying it had resulted from Allende's own disastrous economic policy.

Much as was the case with our Vietnam interrogation, my approach here was to play the devil's advocate, a role I welcomed. By the time these interviews occurred, Chileans were in the grip of a murderous dictatorship, presided over by General Augusto Pinochet, one that would endure until the late 1980s, when Pinochet—mistaking the deference paid the man who carries the gun for public adulation—permitted a free election to take place. Chileans promptly voted him out of office.

FROST: One of the most controversial aspects of your foreign policy, namely Chile. And there were the Chilean elections in September 1970, and after those elections of September the fourth popular tradition suggested that Allende the Marxist would be elected as president in a joint session on October the twenty-fourth. And in a meeting with the CIA director, Richard Helms, on September the fifteenth, 1970, you did direct him, didn't you, to take such steps as were necessary to prevent Allende from coming to power?

NIXON: Yes, there was such a meeting, ah . . . we did not discuss the specific steps, the only steps that were discussed, ah . . . was the use of economic measures that might be effective. And in listing whatever political groups were to be involved, ah . . . in preventing Allende from coming to power, enlisting of course the major military leaders. Because, of course, the military had great influence in Chile, on a political situation.

FROST: But if—

NIXON: A military coup however, was not what was contemplated, and of course it did not take place.

FROST: No, but a coup was one of the things that Mr. Helms could have felt was certainly not ruled out?

NIXON: Well, let's understand what a coup would have been in this instance, ah . . . first we talk about the election. This was a case where as so often happens in the non-Communist world, the non-Communists fight among each other. The Communists are always united. Virtually always. Now, in this case, Allende got thirty-six percent of the vote. Ah . . . the very conservative candidate got thirty-five percent of the vote and the third-party candidate, ah . . . the party of former President Frei, got the rest of the vote. So, as far as the people of Chile were concerned, as far as the vote was concerned, ah . . . the individual who had received a little over a third of the vote, Allende, was going to come to power unless the other two parties would get together and, in the Congress of course, vote for maybe a neutral candidate or some third candidate . . . because the other two parties unfortunately probably disliked each other or fought each other more than they fought Allende, which is the great tragedy in the Western, or should I say, the non-Communist, world today.

So that was the situation we were confronted with. What I anticipated was that, ah . . . it might be possible under the circumstances since Allende had not gotten, ah . . . a majority of the popular vote, that the other two parties should get together and, ah . . . with proper press support and, ah . . . support from the military that they would be able, through a coup, if you want to use that term, to prevent him from coming to power. There was no discussion, of course, of "coup" in terms of a . . . military operation . . . of sending in . . . which would require the huge commitment of arms and so forth. We also have to understand that that was not a new policy as far as the United States is concerned. Back in 1964, ah . . . we spent over four million dollars in the election campaign itself, to defeat Allende and, ah . . . it was generally considered to be a very good move on our part to do so at that time by some who became critics of our attempts to keep him from coming to power this time.

FROST: But nevertheless, you wanted to prevent him from coming to power and there had obviously been lots of cases in democracies of where people have come to power by a vote, well, in your own case the first time it was forty-three percent—

NIXON: Yes . . . and, ah . . . John Kennedy had a forty-nine percent because of the third-party candidate in 1960.

FROST: Right.

NIXON: That's quite true.

FROST: The unique thing here really was that—

NIXON: —neither of us was running against a Communist.

FROST: No, indeed . . . ah . . . but the unique thing here, isn't it that, can you think of any other example where the United States in . . . recent United States history . . . attempted to interrupt the constitutional processes of a democratic government?

NIXON: Well, it depends on what you mean by recent . . . ah . . . well . . . you mean the last four or five years? No, I can't think of any.

FROST: But given that it hasn't happened in recent years and that Chile has a strong democratic tradition and so on . . . and Chile's a small country, does that mean that you think that today, if France or Italy was about to vote a Marxist into power, that similar steps should be taken to try and prevent it?

NIXON: I think in the case of France and Italy today, ah . . . and because of the, what I think the . . . totally irresponsible activities of the Church Committee and some of the media in, in its attacks on the CIA, some of the attacks were justified, ah . . . but . . . this total discrediting of the CIA, creating the impression they're a bunch of assassins. That they're running off on their own trying to poison people and this and that and the other thing, you know, some of the horrible examples that have been printed in the press and, ah . . . put on television and testified to and so forth and so on . . . In view of that situation, ah . . . the possibility of a so-called covert activity in a country like France or Italy is out of the question. I would only point out that, ah . . . as far as those major countries are concerned, like France and Italy, I would think that now that they are strong enough, ah . . . with their own traditions and that the non-Communist forces are strong enough, ah . . . that when they look down, frankly, the gun barrel and see the possibility of a Communist minority, well organized, well heeled, ah . . . joining with some splinter party, ah . . . which is on the left but not Communist and taking power, that this will mean a coalition developing among the non-Communist parties, I think this would, I would . . . I would think this would happen in Italy, just trying to be a prophet, and I could be proved to be wrong. And it could happen in France, although the recent indications as I read after the municipal elections, are that it might be very difficult in France for, ah . . . Giscard d'Éstaing and the Gaullists to work together to prevent the Communists-Socialists group . . . if they are in coalition from getting a majority of the vote.

And they have to face the fact that the people of Europe, the people of any country where they have this decision to make, that as far as the U.S. commitment to NATO is concerned, in my view, speaking as a political observer, it will be . . . be . . . impossible to keep up the U.S. commitment and get the funds from the Congress for NATO if the Communists get it. When the Communists get in, the U.S. will get out.

FROST: But at the same time, summing up your feelings, what you are mentioning about covert activities being found out and so on is that if, ah . . . a Marxist takeover looked as though it was around the corner, ah . . . you would regret the fact that we couldn't engage in covert activity in France or Italy, really. Because we'd be found out.

NIXON: Let me say, ah . . . let's get our priorities as far as morality into proper perspective here. Ah . . . what we're really talking about is the real world . . . not the world as I know you want and as I want it. Different as our backgrounds are . . . we would prefer a world in the great Anglo-American tradition in which, ah . . . we have freedom of expression, ah . . . in which there are not covert activities, there are no fears, no repression . . . or if there is, that, when it is punished, et cetera, et cetera, et cetera, it isn't that kind of a world. We live in a world where at the present time the greatest threat to free nations is not from Communist nations, potential aggressor nations, marching over borders, it is not from Communist nations with huge nuclear armaments launching a nuclear strike, but the threat to free nations is through Communist nations, potentially aggressive Communist nations like the Soviet Union, like Cuba, for example, like Chile if Allende had stayed in power. Burrowing under a border rather than over a border—

FROST: But—

NIXON: —and supporting, and supporting the Communist Party. Now, let's face it, let's look at the situation in Italy. Let us suppose that the Communist Party of the Soviet Union and of the other Eastern European countries states . . . ah . . . put in, ah . . . tens of millions of dollars to help the Communist Party in Italy, ah . . . what are the Christian Democrats going to do? Or the other parties going to do? Can they get any help from the outside? And the answer is, unless the free world develops some method to deal with this kind of covert support by an outside power of Communist forces within countries, then those countries are going to be taken away . . . one by one.

FROST: What did you have in mind in Chile when you said that you wanted the CIA or you wanted America to make the economy scream? What did you have in mind?

NIXON: Well . . . Chile, of course, is interested in, ah . . . obtaining loans . . . ah . . . from international organizations . . . where we have a vote . . . and I indicated that, ah . . . wherever we had a vote . . . where Chile was involved that, ah . . . unless there were strong considerations on the other side, that we would vote against them. Ah . . . I felt that as far as Chile was concerned, since they were expropriating American property—they expropriated, Allende did, it took him only three years to expropriate two hundred and seventy-five firms. I know that, ah . . . as far as—

FROST: He hadn't done that on September the fifteenth—

NIXON: Yeah . . . I know . . . but I knew that was coming . . . all you had to do was to read his campaign speeches. Let us . . . we forget when we talk about Allende that his history went back, we knew him in 1964 when both the Kennedy and Johnson administrations spent a total of over four million dollars to keep him from coming to power.

FROST: Why—

NIXON: Because they knew that he was a Marxist and they knew what he would do to Chile and the effect that would have on the countries neighboring on Latin America. But in his campaign, early in 1970, he said, ah . . . with Cuba in the Caribbean and with Chile on the southern cone, we—he meant Castro and Allende—will make the revolution in Latin America. Now, we had fair warning of that, now, why was Chile, even though it is a small country in terms of population, it has common borders with Argentina, ah . . . it can have influence in Bolivia . . . it can have an influence in Brazil . . . it can have one also in Peru . . . and all of those countries had significant problems and all of them were concerned about the possibility of a beachhead of communism in Chile that would export in the Western

Hemisphere, Cuba, that is, exporting revolution and didn't want another one—Chile—doing it—

FROST: But there are two things there, surely . . . one is that Cuba, which everybody would say is Communist in the traditional sense of the word, Cuba has been totally unsuccessful in its export of revolution or anything else since 1958 and Allende just didn't turn out that way. He turned out to be a Marxist . . . he worked within the system for the three years. He never attempted to introduce political pressure. That only came later. Ah . . . he continued to work within the system to the extent where it was predicted he would lose the next election.

NIXON: Well, as a matter of fact. Let's well understand, ah . . . Allende played it very clever, but he . . . he . . . played it as a Chilean would rather than as a Cuban would. The Chileans being, ah . . . frankly less volatile than the Cubans, I would say, but on the other hand . . . there wasn't any question about his turning all the screws that he possibly could, ah . . . in the direction of making Chile a Marxist state. There wasn't any question but that he was cooperating with Castro. There wasn't any question that Chile was being used by some of Castro's agents as a base to export terrorism into Argentina, to Bolivia, to Brazil. We knew all of that.

FROST: But he never—

NIXON: And also, now, as far as repression is concerned, after all, he did that in a subtle way. But we would have called it repression if it had been done in Vietnam. We would have called it repression if it had been done among any one of our friendly countries today. Ah . . . for example the government-owned television station. The only program that they had was simply Marxist philosophy . . . as far as government advertising was concerned, he squeezed the press . . . in addition to, ah . . . shutting down the UP office or UPI office for a time 'cause he didn't like one of their stories . . . and shutting down *El Mercurio,* the biggest paper . . . first ordering it for a week and then, because of an outcry, cutting it down to a day, because he didn't like some of their stories . . . But he also had a very subtle power in the sense that there the government advertises in the paper and so, he proceeded to do everything that he could to cut that advertising down and that was why the United States in that period, ah . . . very properly, I believe, and President Carter, I think, has shown right judgment in indicating that there was nothing illegal about this and, as a matter of fact, therefore had indirectly approved it in that period—

FROST: We . . . indirectly—

NIXON: —in that period . . . we helped subsidize the Chilean press in order to keep it free, to keep it alive.

FROST: You mean President Carter's indirectly proved that—

NIXON: No . . . he just said that he has checked the record and he found that there was nothing illegal in any activities as far as Chile was concerned.

FROST: But when you look at that list of . . . I mean . . . them closing the UPI office—

NIXON: I don't mean that he said that he would have done it . . . but he said there was nothing illegal—

FROST: When you look at the closing of the UPI office, for instance, and things like that . . . all of those things are trivial compared with what followed Allende . . . I mean, Allende, with all of that list, looks like a saint compared with the repression of Pinochet . . .

This moment presented Nixon with the opportunity to offer his traditional "realist" spin on U.S. interests with smaller states. Simply stated, it was to support the stability of right-wing dictatorships because they lack the internationalist design of left-wing governments. It is a view that shaped U.S. policy toward the Third World throughout the Cold War and for which we have yet to devise a workable substitute.

NIXON: That's right . . . I am not here to defend and will not defend repression by any government . . . be it a friend of the United States or one that is opposed to the United States. But I would make this comparison . . . as far as Allende was concerned . . . he was an opponent of the United States, he also was an opponent of the non-Communist governments who were his neighbors. He was an ally of Castro, ah . . . in other words, we've got to separate our views with regard to what kind of dictatorships are, not that one dictator . . . any dictatorship is good, but in terms of . . . national security . . . in terms of our own self-interest, the right-wing dictatorship, ah . . . if it is not exporting its revolution, if it is not interfering with its neighbors, if it is not taking action directed against the United States, it is therefore of no security concern to us. It is of a human rights concern. A left-wing dictatorship, on the other hand, we find that they do engage in trying to export their subversion to other countries . . . and that does involve our security interests.

FROST: In fact, what they have now with Pinochet is a right-wing dictatorship . . . what they had with Allende was a left-wing, or Marxist, democracy . . . it was never a dictatorship.

NIXON: Let's understand—

FROST: Was it . . . was it, though?

NIXON: No . . . I don't agree with your assertion whatsoever . . . I, oh . . . I would—

FROST: It was not a dictatorship, was it?

NIXON: It was . . . you said it was not . . . a dictatorship . . . and my point is Allende was a very subtle and a very clever man. But he was . . . it was not a dictatorship in, ah . . . the sense that Castro's Cuba is a dictatorship . . . ah . . . it was not a dictatorship in that sense, certainly. On the other hand, as far as the situation in Chile was concerned, he was engaging in dictatorial actions, which eventually would have allowed him to impose a dictatorship. That was his goal . . .

FROST: But the . . . CIA reported shortly before his death that he was no threat to democracy. He wasn't planning to abolish democracy, and he was going to lose in the next election.

This question presented Nixon the opportunity to slam the CIA, something he seemed to do with great gusto. Had the Agency performed as poorly as Nixon suggested? Or was Nixon bitter over the Agency's failure to do his Watergate bidding and ask the FBI to halt its investigation?

NIXON: Based on the CIA's record of accuracy in their reports, I would take all that with a grain of salt . . . ah . . . they didn't even predict that he was going to win this time. They didn't predict that was going to happen in Cambodia . . . they didn't even predict that there was going to be, ah . . . the Yom Kippur War. As far as the CIA is concerned, at that point—and now I understand it is being improved, and I trust it will be under the new leadership—at that point its intelligence estimates were not very good on Latin America. I also go back to the point that in terms of, ah . . . what we really had here in Chile, I think it was graphically described to me, even though you and many of our audience may disagree with what we did in Chile . . . and disagree with my defense of it . . . but I have to state what I believe, and that's what we're here for in this program . . . and I'm going to continue to do it. I don't care whether it's popular or not . . . we must have in mind the fact that what is not popular today maybe has to be made popular if we're to survive tomorrow. And that's what I'm talking about. Because here is what was involved in Chile . . . I remember months before he even came to power in 1970, that when it was thought that he might run again . . . an Italian businessman came to call on me in the Oval Office and said, "If Allende should win the election in Chile and then you have Castro in Cuba, what you will in effect have in Latin America is a red sandwich. And eventually it will all be red." And that's what we confronted.

FROST: But that's madness of him to say that. I mean, how—

NIXON: It isn't madness at all . . . it shows somebody saying, cutting through the hypocritical double standards of those who can see all the dangers on the right—

FROST: No . . . no . . . but surely . . . no, but I—

NIXON: —and don't look at the dangers on the left.

FROST: No, but surely, Mr. President, there's two . . . you've got little Cuba and little Chile . . . and all those enormous countries in between . . . it's like . . . if it's a red sandwich, it's got two pieces of bread here and here and an enormous bit of beef in the middle. I mean, are you really saying that Brazil should feel itself surrounded by Cuba and Chile?

NIXON: All that I can say is that, as far as Brazil is concerned, ah . . . as far as Argentina is concerned, the other countries in that part of the hemisphere, ah . . . I have visited most of them . . . in 1958, for example . . . ah . . . and I can testify to the fact that many of their governments are potentially unstable. Ah . . . I can testify to the fact that also they do have a problem of subversion. They do . . . they fear it . . . maybe their fears are greater than they should be, but . . . what we are concerned about in Latin America, and you have to be concerned, is the instability of so many of the governments there. Oh, I'm not suggesting that Chile's going to launch an armed invasion of Argentina or of Brazil, for example . . . of course they'd get clobbered. And I'm not suggesting that Cuba is going to take an amphibious force over and take over Venezuela.

But I do know this . . . Castro has caused plenty of problems to his neighbors . . . and is continuing to cause problems, and Chile will continue to. I will not take your assumption that it was madness . . . a madman, in effect, or that it was madness for him to even suggest this. Well, he's mad like a fox, because what he's doing is taking the historical view . . . and that is, he knows the nature of communism, the threat, I say again, is not the . . . even the size of the country. It's the fact that they have the beachhead. It's the tactics they use. And in this instance . . . it's having a base to infiltrate from outside. Sending people under borders and over borders and so forth . . . I didn't mean that it was an immediate threat, but I meant that if you let one go, you're going to have some problems with others . . . so we'll just let the red sandwich sit right there because . . . obviously you've got other subjects to cover.

FROST: Well, we'll stick to this one.

NIXON: I . . . oh, I can take this one as long as you want.

FROST: But the point is, really, what I really wanted to say was that Allende won the election, as you said, albeit with thirty-six percent . . . we wanted to prevent him from coming to power. As it turned out, whatever may have lurked in the secret heart of a Marxist working within the democratic system, ah . . . he hadn't abolished the democratic system . . . he hadn't abolished the principal election or anything else, and that was abolished by Pinochet. Now, in retrospect, looking back, don't you think the Chileans have had this tradition of democracy going back to 1818, with three temporary interruptions in between . . . nothing interrupting it since 1932, till Pinochet, in retrospect, don't you think that the Chileans were a better judge of what would preserve their democracy than you were?

NIXON: Who do you think overthrew Allende? You see, there's the point you've missed. Allende, at the time he had been in charge of government with his programs that destroyed agriculture and discouraging foreign investment, ah . . . because obviously, I wasn't going to approve any American loans to companies to invest in Chile when it might be expropriated. That was one of the economic squeezes we put on them—

FROST: The main reason—

NIXON: —but in any event . . . well, let's look at what happened. We forget Allende was thrown out of office not by any outside coup, he was assassinated . . . his assassination took place or his death, whether it was assassination or suicide . . . there seemed to be some argument there . . . as it was in the case of Diem, but be that as it may, he was a casualty. As far as Allende was concerned, it had ruined the Chilean economy. Inflation was three hundred percent. It took him only three years to take Chile, which had had a pretty good record up to that time in meeting its international obligations . . . to make Chile one of the poorest credit risks in the whole world. Ah . . . now, under the circumstances, with that kind of a record, it was the Chilean people, ah . . . the Chilean leadership that supported the people . . . even the women marched in the street . . . and so forth . . . the labor unions turned against him . . . they overthrew Allende.

I come back to the proposition that I made previously, ah . . . and, ah . . . as far as the countries that have Communist governments are concerned, and this is true of those in Eastern Europe, it's true of Cuba, ah . . . I think it will eventually prove in Angola . . . although we don't have any word out of it yet . . . it will be true in other parts of the world as well . . . ah . . . but as far as popular support . . . I would have to say that Foster Dulles was right . . . that the great failure of communism is that they seem irresistible in their ability to conquer a country either over or under a border, but they are totally inadequate and always fail in winning the support of the people of the countries they take over. Allende lost eventually; Allende was overthrown eventually . . . not because of anything that was done from the outside . . . but because his system didn't work in Chile . . . and Chile decided to throw it out. Now, what they have now, I don't approve of the repression, I would hope that the Chilean people, Chilean leadership, would change their policies. But I can only suggest this . . . neither the British nor the French nor the Americans, ah . . . can any longer take the position that our form of democracy—and each of it differs, the French have a different one from the British, the British a different one from the Americans, and the Italians have a different one from anybody else. But the point is, our form of democracy . . . our form of freedom will not work in many other countries in the world.

FROST: But on that principle . . . surely we shouldn't have tried to interfere in the process in Chile. I mean, we should surely believe in self-determination in the sense we went to war for it in South Vietnam and we said indeed we'd accept whatever government they elected even if it was a Communist one, much less a Marxist one. Ah . . . surely we should have done the same thing for self-determination in Chile. I mean, how do you decide which results in a democracy you don't accept. I mean, what are the rules? We tried to prevent Allende from coming to power . . . (who'd won) because we didn't like the result. We should not try to impose our will on other countries . . . you said, therefore, we shouldn't have tried to impose it on Chile.

NIXON: Well, it's obvious that, ah . . . we're in an area where we have a very spirited and friendly disagreement. Because we're both on the same side. Ah . . . we both want freedom of choice for all the world. We would both like a situation where there is no discrimination in voting. We would both like a situation where there was freedom of press and freedom of religion all over the world. We would both like a situation where everything was perfect as it is in Britain, as it is in America. And we know very well it isn't perfect either place, and it's never going to be because men are men and women are women and none of us are perfect . . . but the point that I'm making in the case of Chile and what distinguishes it from the other is this . . . that when, as was the case in Chile, you have a situation where an outside force, the Cubans, put three hundred and fifty thousand dollars at a minimum into the election in 1970—that got Allende into power. The Soviet Union put in what is called by the CIA's report an amount that is unknown. I would imagine it was in the matter of millions of dollars, because when they play, they play hardball. And when you have a situation where a government comes into power supported by an outside power, a Communist power, ah . . . then I believe that there is justification for supporting those . . . seeing that those who are not Communist and who are in the majority at least have an equal show at the ballot box. Ah . . . and second . . . that, ah . . . once the election process . . . electoral process takes place, to keep their right to dissent alive by supporting their newspapers and their radio stations and so forth.

FROST: But they hadn't had—

NIXON: Because what we had is—

FROST: —equal shot at the ballot box?

NIXON: Yeah . . . it wasn't an equal shot, because basically when you have, in this instance, when you had Allende, with massive support from outside of Chile by Communists, and because a division within the non-Communist forces in Chile and perhaps the United States not recognizing how serious the problem was . . . very little support from the United States on the other side . . . it wasn't that fair a fight.

FROST: But the three hundred and fifty thousand from Cuba is just equivalent to the sum from IT&T, and we spent that eight million dollars over that three-year period in—

NIXON: Oh, now we're talking about this period—

FROST: —in the period after—

NIXON: —the election . . . I'm talking about the period before the election—

FROST: But we support the non—

NIXON: —before the election, and I should also point out that in that instance you're leaving out how much the Soviet Union put in.

FROST: Do you think that over these years the Soviet Union . . . does your reading of reports, et cetera, et cetera, suggest that they put in as much as the United States?

NIXON: Where?

FROST: Into Chile.

NIXON: I don't know how much they put in, but my point is, they put in enough that they, they, ah . . . I would certainly say matched us dollar for dollar and probably did . . . it was used better due to the fact of the division, as I said, among the non-Communist forces.

FROST: If you had to choose a word to describe the Pinochet regime, what adjective would you use . . . *brutal?*

NIXON: Well, when they are brutal, yes, ah . . . when they are dictatorial, I would say they are dictatorial. Ah . . . I would also have to, on the other side, indicate that they are non-Communists and that they are not enemies of the United States and that they do not threaten any of their neighbors. Now, basically, what we have to understand—and this does not justify a right-wing dictatorship or any kind of dictatorship—but what we have to understand is that, in this instance, the present Chilean government is engaging in activities that we disapprove of in terms of its internal policies. But as far as its external policies are concerned, they don't threaten any of their neighbors and they don't threaten us.

10
VIETNAM

FROST: Mr. President, the whole area of foreign policy is such a vast one, but at the moment you took office, America's involvement in Vietnam was regarded by many as a disaster that was splitting American society at home in a very grievous way, for what seemed to many an obscure or even a mistaken reason. How did it look to you, though?

NIXON: Well, it looked to me first that, ah, the reason for our being in Vietnam had perhaps not been adequately understood by the American people. I understood it because I had been there on many occasions, ah . . . my experience, as a matter of fact, went clear back to 1953, when I visited Hanoi, at a time before the Communists took over in '54 in that area, Saigon as well.

On the other hand, in 1969, when I took over, I'll never forget, I went into the bathroom, which President Johnson had shown me when he took me around the White House, and there in the president's bathroom was a wall safe built into the wall. You can't tell that it's a safe, you have to hit it and then it opens, and it's where the secret documents are kept when a president takes reading material over to the residence that is classified. And so, that first night in the White House, after the inauguration parade and all the rest, I popped the door open and, ah . . . there was only one piece of paper in the safe; it was the president's briefing for January the tenth, 1969. And, as I looked at that briefing, the whole of Vietnam came into perspective. Ah, almost three hundred a week had been killed, the year of 1968, President Johnson's last year in office. And the rate was at about three hundred even then. There was no progress in the Paris peace talks, which had begun during the election, as you may recall. He announced it on October 31. The situation, in terms of the military action, was very difficult from the standpoint of the United States in another way, because he had, at that time, 538,000 Americans in Vietnam, ah . . . 14,000 a month were being drafted. That was Vietnam then.

So, from my standpoint and looking at it just from a political standpoint, I must say that some of my friends, politically, who strongly urged me "Just get out of Vietnam," and, ah . . . "Put the blame on Johnson and Kennedy, who got us in," made political sense. It didn't make sense for the country, in my view. It didn't make sense for the world, which is more important. And, needless to say, of course, it didn't make sense for the people of Vietnam, although I recognize that perhaps most people in America aren't too concerned about what's happened to those people since the Communists have taken over.

But at that point in time, we had to recognize, too, as I read that morning report, the last one Johnson saw and the first one I saw, we had no communication whatsoever with the People's Republic of China, and they were helping the North Vietnamese. Ah, our relations with the

Soviet Union were virtually at a standstill, despite indications that they might be of assistance in working toward a peace agreement in Vietnam. They had actually not done so after the peace conference started in Paris on October 31. Our European allies, I knew from having traveled there in 1967, ah, had great concern about the reliability of the United States, ah . . . as an ally, because they thought we were bogged down in Vietnam, and they also wondered if, in the event something happened there, the Americans, after going through Vietnam and Korea, two very difficult wars, ah . . . ah . . . might keep its commitments. And they raised those questions with me.

Ah, the situation in the Mideast, while the war was two years past, the, what is called the June War of '67, I was there right at the end of that war, was still terribly difficult, with the United States having no relations, for example, with Egypt, and no relations with Syria, of course, a close relationship with Israel. Ah . . . but it was there, an explosive area, and consequently, the Israelis—as my discussion with Ambassador Rabin, who saw me shortly before the election, indicated—the Israelis were concerned about whether the United States' position in Vietnam . . . if we decided to bug out, might mean that if Israel faced an attack from its neighbors, supported by a superpower like the Soviet Union, whether the United States would keep its commitments. No, what I've tried to give you, basically, is an overview as to why I rejected, out of hand, the political argument: "Well, Kennedy sent in the first 17,000 men," which he did, "and Johnson, of course, sent in the rest," so that was 530,000.

Throughout our discussion of Vietnam, Nixon presented himself as a man who had inherited a brutal, unpopular war and who made a number of close, unpopular, courageous, and correct decisions. My goal, throughout this portion of the interview, was to remind both Nixon and the viewing public that there had been alternatives and that Nixon's decision to proceed with the war had carried some enormous costs. These twin themes recur throughout our discussion of Vietnam, from Nixon's first efforts at linkage to his final wrath over Congress's failure to prevent the North Vietnamese victory.

NIXON: I could have said, "This was a war that I inherited, we shouldn't have gotten in in the first place, and I'm getting out, and let the Communists take over." I didn't do it. And I didn't do it for two reasons. Two major reasons, I would say: one, because I think John Kennedy was right when, as late as July 1963, he said that "if Vietnam were to fall, all of South East Asia would fall with it." And he said, "We are going to stay there." I think Lyndon Johnson was right when, very late, a little too late, as a matter of fact, finally, in Baltimore, he made a speech in early '68 in which he pointed out the reasons why, that what was involved in Vietnam was not simply freedom for Vietnam, the chance to choose their own form of government, but the security of the United States, the credibility of the United States as a dependable ally, and, as far as our adversaries are concerned, as one that would take positive action to stop aggression. So under those circumstances, I thought, first, that Kennedy and Johnson were right in going into Vietnam. I was very critical of the way the war had been conducted. I thought they could have done, particularly President Johnson, because, of course, he had the major responsibility—we were in deeper by the time he was president—that they could have conducted it in a more effective way. I had some ideas as to what could be done, but I wasn't about to go down that easy political path of bugging out, blaming it on my predecessors. It would have been enormously popular in America, but that would have paid . . . have been an enormous cost, eventually, even to America but particularly to the whole free world.

Nixon had outlined one set of costs; it was my job to remind him of the other.

FROST: But wasn't staying there, I mean, that was also at a massive cost, wasn't it? In billions of dollars; 138,000 South Vietnamese killed, half a million Cambodians, half a million North Vietnamese, and so on? That cost . . . it's a question of weighing one cost against another cost, isn't it? But you thought that cost was worth paying for what you got?

NIXON: Looking at my term in office, yes. I think, considering the kind of peace agreement we finally got in January of 1973, one which provided for a cease-fire, ah . . . one which provided for, of course, the exchange in return of our POWs. One which also provided for no violations in the future of South Vietnam's territory by the North Vietnamese, among many other things. I believe that having accomplished that, after those four long, tortuous years, was worthwhile. And that held for over two years.

The cost, I agree with you, however, was very great. It was a close call. A very difficult call. It, it was difficult for me personally because, ah . . . you know, one of the problems a president has, and President Johnson once reflected on this point the first time I saw him after I was nominated, before I was elected, and this was the question as to whether there would be a bombing halt. He was violently against it. And I remember his leaning across the table and pounding the table when he said, "How can I tell that boy that's being shot at in Vietnam that I'm going to stop the bombing in North Vietnam when they're killin' him or that bullet's gonna kill him?" He says, "I'm not gonna stop the bombing until I get some kind of understanding from 'em that I know they're going to keep." No, as far as I was concerned, I felt just as deeply because a president has to write the letters to the next of kin. A president also has the responsibility of looking over casualty figures every day. A president, of course, knows what is happening to the people of South Vietnam, the people of North Vietnam. He knows the cost of war. And also, I knew, as a political man, the terrible costs at home. I know that not only the Democratic Party was so divided that they drove President Johnson out of office.

I knew there were many Republicans in the House and in the Senate, a minority but a considerable number at that, who felt that we shouldn't have gotten in in the first place, and there were some, who for the wrong reason, of course, as I've earlier indicated, that whether we should have gotten in or shouldn't have gotten in was irrelevant. That, now, that the war wasn't popular, we should just, I should take the political easy way out. And I didn't take it for the reasons I've mentioned, and I think that decision was right.

In a famous interview with Walter Cronkite, at a time when only sixteen thousand American troops were in Vietnam, President John F. Kennedy endorsed the domino theory as reflecting his own analysis. Did Nixon harbor the same belief? If so, I suspected that many would conclude that his decision to pursue the war was based on what by 1977 was a discredited rationale.

FROST: Right. Well, let's try and analyze through the story of those years, and just to clarify one thing going in, really, your prime motivation was American credibility, did you still believe in '68 the domino theory?

NIXON: Oh, I certainly did. I knew that there are those who say that nothing, ah . . . can be proved in terms of the domino theory, but it's only a question of time. It's a question of a . . . as I have often said, ah . . . those who say that they, there will be no domino effect haven't talked with the dominos. I have talked to the leaders in the Philippines. I have talked to the leaders in Singapore and Malaysia, in Thailand. Thailand and the Philippines, of course, being our allies. I've talked to Suharto in Indonesia, all of this in 1967. And I can assure you that every one of them were virtually terrified at the thought that the United States might bug out of Vietnam, because they felt that would be an indication that they, particularly those that were allies, and in the case of Indonesia, one that considered that it was a friend of the United States and had really cast their lots with us, they were concerned if we wouldn't stand up after all we'd done in Vietnam, if we bugged out, could they depend upon us there? And they thought that tide of aggression would go on. You see, the major point they were concerned about was not just saving the people of South Vietnam, they were concerned about saving themselves. And they were concerned that the same kind of aggression that was being waged against Vietnam by the North Vietnamese, with the support of the Soviet Union and, at that point, to a less extent by, by the PRC, the People's Republic of China, that same kind of aggression, if it were not stopped in Vietnam, would then be visited upon them and that if the United States bugged out of Vietnam, we wouldn't stand up for them. And I think they were right.

FROST: But the reading of the Pentagon Papers suggests that most of the key Johnson officials had given up the domino theory by '67, and they were really talking about American credibility solely, weren't they?

NIXON: Well, if you read the Pentagon Papers, ah . . . what you're reading are the options of people who have the great luxury of not having to live with their decision. Ah . . . of not having to see the whole picture. I don't presume to be a great expert in foreign policy or a genius or anything of that sort, but I learned it the hard way and, I believe, the most effective way. I had been there. I had been in Vietnam a half a dozen times. I have been in the countries surrounding Vietnam, and incidentally, let me say, Japan was vitally interested in what happened in Vietnam. Oh, they didn't say it publicly, but the Japanese, their greatest trading area, except for the United States, was Southeast Asia. They were interested in it economically. And, too, they were concerned about their security, because the Japanese, as you know, are an economic giant and a military pygmy. And, under the circumstances, they rely totally on the United States. When we talk about credibility, credibility for the United States, that isn't said in terms of jingoism or anything like that, it's said in terms of whether or not those in the free world who are allies of the United States or who, like Israel, are not joined with us in an alliance but depend upon the United States, whether or not they can count on the United States in the event they are threatened with aggression, either over a border or under a border, supported by a major power. And that's what Vietnam was about to them, and they understood it.

I decided to question Nixon about a rumor circulated by the old China Lobby to the effect that Nixon had urged the group to turn thumbs down on Lyndon Johnson's October halt to the bombing of North Vietnam. In order to discredit the president's election-eve gesture, Nixon had reportedly pledged a tougher line against Hanoi should he, rather than Hubert Humphrey, be elected president. The incident had surprisingly little in the way of political legs, as it was denied by Nixon, who was promptly elected president. But it does show the bitterness of the China Lobby toward its former friend and benefactor, and it is similar to the "October surprise" Democrats would associate with Ronald Reagan during the future Iranian hostage crisis.

FROST: One little brief, rather bewildering footnote to history, as we come into this whole account on which new material has been appearing recently, you mentioned the nearby Paris peace talks just before the election, which you won in '68, and there have been a lot of reports that Madame Chennault was in touch with President Thieu via the Vietnamese Embassy and urging him to take a firm line, because he would get better . . . better terms, better support from you than from a Democratic president, and so on. Did you ever hear about that?

NIXON: I, of course, I'm hearing about it from you today, and I have read the reports. Just let me make this one point very clear, however. As far as I was concerned, as I told President Johnson on the phone when he informed me on October the thirty-first that he was going forward with the bombing halt, and when he said, in answer to my direct question, that he had solid commitments that they would negotiate seriously, that they would respect the DMZ and that they would quit or reduce their shelling of the cities, the major cities of South Vietnam, he said, I said, under the circumstances, I would support the talks. Let me, and I would do nothing to undercut them. I did nothing to undercut them. Ah, as far as Madame Chennault or any number of other people, like whoever may have felt, if they did, that the talks should not go forward or that the South Vietnamese should not go along. I did not authorize them, and I had no knowledge of any contact with the South Vietnamese at that point, urging them not to do so. Because I couldn't have done that in conscience. I wanted the talks to succeed if possible, because, frankly, I expected to win the election, although it was becoming close, and to have the war behind me rather than having it on my plate with talks that had failed would have been to me a very great advantage.

FROST: So that when Madame Chennault says that she stated recently, "From the very first conversation, I made it clear I was speaking for Mr. Nixon, and it's clear the ambassador was only relaying messages between Mr. Nixon and Mr. Thieu." Ah, she's not telling the truth?

NIXON: Whatever she has said, she has said. What I am telling you is the truth, as far as I, ah . . . ah . . . as far as my position was concerned. Throughout, I was briefed by President Johnson, as was Vice President Humphrey, my major opponent, and, ah . . . Governor Wallace. Because he'd always get us on a conference call, the three of us together, ah . . . when he informed of the developments. As far as I was concerned, I played a very responsible line. Incidentally, President Johnson rather ruefully told me, about several months later, when I was flying with him out to California to dedicate a, ah . . . grove of our great redwoods to Mrs. Johnson, who, as you know, had worked on the beautification program, he told me ruefully, he said, "You know, Hubert gave me a lot more trouble in those weeks before the bombing halt than you did." Now, what he was referring to, of course, was that Hubert Humphrey, on September 29, had indicated that he, if he were elected, ah . . . would stop the bombing, provided that the, ah . . . ah . . . North Vietnamese, ah, would respect the DMZ. Well, now, Johnson had already been insisting on far more than that, and so, consequently, it made the North Vietnamese raise their price if they thought he was going in. As far as I was concerned, however, ah . . . I had no influence on Johnson and I certainly don't believe I had any influence on Thieu, and as far as contacts with Thieu were concerned, I should point out, because I know your research may have probably over . . . possibly overlooked this or you wouldn't have asked the question in the way that you did, President Johnson pointed out in his memoirs that, before he made his announcement on October 31, that he learned that Thieu was not going to go along and he said, quote, "We therefore have to go it alone." So, as far as President Thieu is concerned, he apparently needed no urging.

During the 1968 campaign, Richard Nixon was widely credited, or, alternatively, accused of harboring a "secret plan" to end the war in Vietnam. So widely discussed was the matter that my researchers were astonished when their work indicated that Nixon had neither bragged of having such a plan nor even claimed to have one. What Nixon did bring to office was a willingness to push various buttons in the hope that one would start the ignition and lead to a negotiated peace.

FROST: Now, what was the first thing, or the first things, that you did to try and solve the problem with the Vietnam War?

You'd mentioned the plan in '68, but now you're in office, in early '69. Were the first steps, the major steps you took, steps involving Russia or what?

Nixon was grudging whenever we discussed concessions he had made to the North Vietnamese, but the following pages suggest that there were many. A combination of Vietnamization, a strictly pursued reduction of U.S. combat troops, and an in-place cease-fire allowing the North Vietnamese to put new forces into the field would contribute to the conviction of many that what Nixon achieved was more in the nature of a fig leaf than peace with honor.

NIXON: Well, first, I didn't mention secret plans in '68. Ah, that was, of course, a press distortion, ah, one of those understandable ones that, ah, you as, know sometimes, your colleagues indulge in. Ah, what I had said was that we needed a new program for Vietnam. One that would end the war and win the peace. Now, that was a very carefully crafted statement. I know from World War I, from World War II, from Korea, et cetera, that ending a war can be only the beginning of trouble. It may lead to another war unless it ends in the right way. And there, for example, I could have ended the war in Vietnam by bugging out, but what would, in my opinion, have not . . . would have lost any chance for peace in the Southeast Asian area at that particular time when those nations didn't have the capacities for self-defense they even have today.

Then, in my, I think, first press conference on January the twenty-seventh, the week after I was in office, I indicated our position, ah . . . we changed the position, trying to make a breakthrough, ah . . . because in this particular instance, what Johnson had announced was that we would halt the bombing in return for stopping the shelling of the cities and respecting the DMZ. We decided we needed a new approach. And the new approach that we developed was that we would have a mutual withdrawal of all American forces and all foreign forces, including, of course, North Vietnamese forces, within twelve months of an agreement and supervised elections and a supervised cease-fire. That was our first step. Then we came on, as the months wore on, ah . . . there was no progress in Paris on that. They turned thumbs down. Their attitude was whether it was a question of first stopping the bombing, which they said might bring progress in the negotiations. And then they said we had to have something with regard to withdrawal. Now, we'd offered mutual withdrawal. No, they said, "That isn't enough." And so we got to the bottom line with them very early. A line they hung to right into the last, until October the eighth, 1972. Whatever we did, mutual withdrawal, unilateral withdrawal, ah . . . nothing that we offered would they consider unless we agreed, on our part, to overthrow the government of South Vietnam and allow them to take over. And that we would not agree to.

Nixon's early efforts to end the war were three-pronged. First, linkage: try to get the Russians and the Chinese to lean on their Hanoi friends to accept a negotiated solution that stopped short of outright U.S. capitulation. Second, increase military pressure—for example, by widening the bombing of infiltration routes that passed through Cambodia. And third, "Vietnamization"—gradually handing all combat responsibility over to the South Vietnamese. None of the three would work. Linkage, for example, failed because in the crunch, détente with the Russians and normalization with the Chinese was more important than Vietnam. Military pressure was something the North Vietnamese had learned to accommodate while battling the French. And Vietnamization became an end in itself, unlinked to any reciprocal North Vietnamese conduct. My questioning sought to flesh this out.

FROST: And when did you first try to enlist the aid of the Soviet Union, in moderating, perhaps, the North Vietnamese position?

NIXON: As far as my ideas as to how we could bring the war to an end, ah . . . we had what I would call a three-track policy. It was one that I had discussed at great length with Dr. Kissinger before I, in my own mind, ah . . . before even telling him, made the decision of December that I would ask him to be my national security adviser. And I found that he went along with this general approach. Certainly, with the great strategy, or the grand strategy, ah . . . although that we might have some tactical differences. One that we should continue to negotiate, and try to find some new formula that would bring the two parties together. To, ah, we should combine that if negotiations reached a dead end or a roadblock, with increased military pressure, because we felt that with increased military pressure, that that might help. Ah, and . . . as far as our complete military pressure was concerned, ah . . . ah . . . increased military pressure was concerned, it was to be combined with training the South Vietnamese forces so that they could take over more of the burden of the war. Because, as Vice President Ky pointed out, ah, several months later, I believe it was, "The Americans have captured our war." And that was the case when we came in.

But the third, the third leg of this three-legged stool, as far as the strategy was concerned, was to go to the heart of the problem. Without Soviet arms, without Chinese support, the North Vietnamese would not have been able to wage a successful war against South Vietnam. So that was the reason, one of the reasons, that I felt it was important at least to start a dialogue with China, although not the major reason. We'll go into that when we have . . . we discuss our China initiative. As far as the Soviet Union is concerned, it was one of the reasons, although not the major reason, that I decided we should have a policy of negotiation, rather than confrontation with them—

FROST: How early—

NIXON: —because I felt that the Soviet Union could play it. Now, getting to your point as to when. I have to lay the background only to point out that the "when" was from the first time Dobrynin, ah . . . I met him after the election, of course, before the inauguration I had known him or met him briefly before, when he was an ambassador, and even when I was out of office and had learned to respect him, ah . . . as a very skilled, perhaps one of the ablest ambassadors on the international scene. Ah, but we had several conversations. I had several conversations with him, and also I set up what we call a private channel, where Dr. Kissinger had conversations with Dobrynin, with regard to what the Soviet Union might do in order to help bring the war in Vietnam to a conclusion, because, as we pointed out to them, as long as they were supporting the North and we were supporting the South, first, there was the danger of the conflict escalating in a way that would draw both of us into confrontation, which neither of us wanted. Second, it cost us, both of us, it cost us more because while it cost them just money and arms, it cost us men, and they knew that, and therefore, I think they were not quite as enthusiastic about the quid pro quo in this instance.

But the other point was that I felt that as long as the war in Vietnam was going on and they were on one side, and we were on the other, that that made it very difficult for us to make progress in other areas, which were of vital interest. For example, it was vitally important that we try and put a tap on the nuclear arms race, what we called the SALT agreement. It was vitally important to do something about the Mideast. And that's why, and this is one of the, I think, one of the major reasons our foreign policy succeeded, we linked, ah . . . agreements, ah . . . negotiations on nuclear arms control, on the Mideast and other areas, to cooperation in trying to settle the problem like Vietnam.

FROST: Now, I can see how linkage operated successfully in certain areas, but why didn't it work with the North Vietnamese? Did the Russians try and influence the North Vietnamese, or were the North Vietnamese genuinely independent of the Russians?

NIXON: Well, the Russians told us that they couldn't influence them. Ah, ah . . . we didn't take that at face value. Ah . . . we couldn't. After all, they could have influenced them by cutting down on the flow of arms to them, and their words, therefore, had a very hollow ring. On the other hand, after our May 8 bombing and mining . . . just before the Russian summit in 1972, then the Russians, I am confident, did use their good offices to bring influence on the North Vietnamese. They were in a difficult position, though, and the same is true of the Chinese. If you were to ask, why wouldn't the Chinese influence them? We have to recognize that the Chinese and the Russians, first, were at odds with each other. Ah . . . and second, they were in competition with each other throughout what is called the Third World as to who was to be the leader of the forces of Communist revolution. Now, the Soviet Union couldn't be caught in the position of being less helpful to a Communist brother, of their Communist brothers, as they called them, in North Vietnam, than the Chinese. And the Chinese couldn't be caught being less helpful than the Soviet Union. And so, while each of them may have had reasons, good reasons in some instances, to want to cool it in Vietnam, because of their desire to have better relations with the U.S., they were always restrained by the fact that they were looking over their shoulder to see what the other might say in the event they got caught taking that kind of action.

So I'm only explaining maybe why they didn't do more than they did. I do know that we did importune them on occasion after occasion, very strongly, for example, in a conversation I had with Dobrynin in October. And I always remembered 1969, where I laid it out chapter and verse to him, that when he came in with complaints about the state of American and Soviet relations from Brezhnev, I, ah . . . said . . . "Well, I have a few statements to make with regard to your attitude toward us." I took a yellow pad out of my desk, and I pushed it across to him. We were on a good relationship, and, but . . . he came in then in a very stern way . . . and a very formal way, and I said, "Well, you'd better make some notes." And the first thing on that notepad, I said to him, "Now, on Vietnam, it's quite clear, that, ah . . . despite all of the assurances that you have given previously, you haven't done anything except continue to support arms, and you've done nothing to help in the negotiations, and we do not consider that can form the basis for our going forward on some of these other areas." We didn't condition it absolutely if . . . ah . . . in the sense of saying, "Look, if you don't help in Vietnam, we're not going to negotiate in other areas." But we did indicate in more subtle and far more effective ways that if they refused to help in Vietnam, it made it difficult for us to be forthcoming in other areas of interest to them. And, that was basically the concept of linkage.

FROST: And, when linkage wasn't working in North Vietnam, that was when Vietnamization became more and more clearly almost a substitute for linkage, was it?

NIXON: Well, as a matter of fact, all of these, all of these policies moved forward together. Ah, they moved in tandem. Ah, it's sometimes in the field of foreign policy and in domestic policy as well. It is such a tendency for the average observer, or even some of the more sophisticated observers in the media, to think in sort of black-and-white terms. Whereas, actually, they are very completely interrelated matters. Our policy in Vietnam, as I indicated, was basically a three-legged, or even, perhaps, if you call Vietnamization, ah . . . bring that into it, a four-legged opposition. Negotiate, military pressure, ah . . . going . . . working on their arms suppliers, the Chinese and the Soviet Union and any other countries, even the Romanians, who might have good relations with them, and then, finally, Vietnamization, providing the Vietnamese with the means of help themselves.

Henry Kissinger reportedly had some doubts about Vietnamization, a concept identified mainly with Defense Secretary Melvin Laird, Nixon's closet dove. Over the next four years, the situation on the ground would fluctuate. Stunning Communist moves would be followed by costly defeats. The South Vietnamese at times fought poorly and other times with great skill and courage. The diplomatic moves resembled a roller-coaster ride, careening wildly from place to place but always ending up where they had begun.

Throughout it all, Vietnamization became an immovable force. By the time the peace negotiations approached their endgame stage, the United States had no combat forces left in Vietnam and arguably little negotiating leverage.

NIXON: We started Vietnamization, I should point out, we started Vietnamization, ah, long before we, ah . . . concluded that we weren't going to get much help from the Soviets, because the first announcement of the withdrawal of forces from Vietnam was made at Midway, as you recall, ah . . . in the middle of 1969. It was only in the fall of 1969 that we began, that we came to realize, that, ah . . . despite our highest hopes, the Soviet Union wasn't going to be particularly helpful.

FROST: There are reports that Vietnamization as a concept, was, most of all, Melvin Laird's idea . . . that Dr. Kissinger had doubts about it. Is that accurate?

NIXON: Well, whether Dr. Kissinger had doubts about it, ah . . . or whether it was Melvin Laird's idea, I would say that you should ask them. I can only tell you my impression of what their views were or what they told me. Ah . . . Vietnamization, ah . . . Mr. Laird was for . . . and he implemented it very effectively. As far as the doubts were concerned, ah . . . ah . . . I would certainly say that Dr. Kissinger was actually taking the lead once he knew that was to be our policy, he was taking the lead in pushing for more progress by our military to train the South Vietnamese so that they could get Americans out. But as to where the concept developed, could I point out that I believe that I announced the major concept. I set it forth in my own, ah . . . speeches, ah . . . before the election, during the election period. And it has been a view of mine for many, many years.

I remember that, for example, that, ah . . . when I saw Yahya Khan, whom you may recall was, ah . . . ah . . . once one of the great prime ministers [*sic*] despite his falling on difficult days toward the end, but he informed me, when I saw him in 1966, he said—I was not then in office—he said, "Mr. Vice President," he said, "you're making a mistake in Vietnam." He said, "You must recognize with Vietnam, with us—Pakistan or any other country—you must help us or them, these countries that cannot defend themselves. Help them fight a war, but don't fight the war for them." And from that time on, and even earlier, but from that time on particularly, I kept preaching that doctrine. I formulated it in what was called the Nixon Doctrine, or the Guam Doctrine. It was announced at Guam. When I said that the policy of the United States would be, in the future, one of helping other people, other nations to help themselves . . . with economic assistance, with military assistance, but to depend upon them to defend themselves, except, of course, where we had a treaty commitment that might require us to come in, we would keep our treaty commitment, or . . . except in those instances where they were threatened by a major nuclear power, where, of course, we could not expect a small country to defend itself.

That was the Nixon Doctrine, and that is Vietnamization. And it should be the policy, and I would hope would continue to be the policy, for the United States in years to come. Because the time when an American president can get the support from the Congress, let alone the American people, to commit American forces to help, to go in and do the fighting in any country, even an allied country, ah . . . ah . . . rather than depending upon that country to defend itself, that time may have passed. No, I'm not suggesting that, as far as our NATO treaty commitment is concerned, that the United States should not, and would not, stand it. It will be difficult, but we would. I would only point out, however, that one of the sad fallouts of what happened in Vietnam is a recent . . . was a Harris Poll, not too recent but about a year ago, in which he asked the question—this is after Vietnam, "In the event of the independence or security of various nations are threatened, would you favor the sending of American forces to defend them?" Do you know what he found? Which was surprising to me—

FROST: No, I don't.

NIXON: Only twenty-two percent, as I recall, would have favored sending American forces to defend Israel. Even if Israel were attacked by all of its neighbors supported by the Soviets. Less than fifty percent, forty-one percent said that they would support going in to defend Berlin, and that, of course, is NATO. The only nation in the world which received over fifty percent was Canada. I don't think he asked about Mexico, perhaps we would have included that as well.

All right, that indicates, of course, a very difficult position for an American president, an American Congress, responsible leaders of a nation in terms of attempting to educate the public and mobilize the American people to, ah . . . on the basic issues. First, that we must keep our treaty commitments in any event. And second, that the way to avoid getting Americans committed is to build up the strength of those nations who are threatened or from without, to defend themselves through adequate military economic assistance. And second, to discourage potential aggressors by having that aggressor have in the back of his mind that the United States just might come in. An indication of what we've, the great lesson of Korea was that once a nation indicates that you will not resist, you invite aggression, and it's that that is the current problem, which I trust the new administration will face up to and I think may face up to very strongly.

FROST: At this point, let's move to Cambodia for a moment, because one of the early decisions of the administration was Operation Menu, the secret bombing of Cambodia. Now, why secret?

NIXON: Cambodia, at that time, in the spring, early spring of 1969, ah . . . great numbers of North Vietnamese troops and great, ah . . . ah . . . amounts of war material: bombs, rockets, et cetera, et cetera. What was happening was that those troops came, ah . . . would come over into Vietnam and then go back. Okay, now, the Joint Chiefs, as the spring wore on, were fearing another Tet Offensive. And they said the only action that we could take in order to blunt it, that they could think of, would be to attack those sanctuaries.

The difficulty was that Cambodia was a neutral country. However, I had noted, at that point, that Sihanouk, whom I met when I was vice president and had met again in 1953 and knew on a personal basis, I'd noted that when he had talked to Chester Bowles in January of 1968, he'd made a very interesting comment about the North Vietnamese–occupied portion of Cambodia; this strip of land which borders on Vietnam. He said, quote, "I can't say this publicly, but I hope that the Americans engage in hot pursuit in bombing in all those areas where the North Vietnamese are and where there are no Cambodians because I don't control that territory." But notice what he said: "I can't say it publicly." Now, as far as the secrecy was concerned, it was imperative.

If we had announced, "We're going to bomb Cambodia," Sihanouk, then, had a problem within his own country, perhaps because there were some who opposed him, would have opposed him, on that issue. He also had a major problem with China. He was very close to the Chinese, the PRC, and we didn't have any significant relations with the PRC at that point. Our initiative hadn't gone very far. So, under the circumstances, it had to be a secret because had it been announced publicly, he would have had to denounce it as being a violation of his neutrality, which, as I will point out or have pointed out, it was not, in my opinion. And, second, it would, of course, have caused him problems with the Chinese, as well as with some of his own people.

FROST: But when it did become public in May, through the piece in *The New York Times,* I mean, he wasn't embarrassed. I mean, he didn't speak out violently against it or anything, as it turned out.

NIXON: No, he didn't at that point, but, on the other hand, he did not have, I should point out, he did not have what would have caused him to speak out: a presidential statement to the effect announcing, "It is our policy now to bomb in Cambodia." Let me under . . . let me, let me say that these are all subtle and close calls. Let me say, too, that we talk about secrecy. I informed the top leaders of the Senate

Armed Services Committee, Dick Russell, John Stennis, about it. They were totally aboard. We, ah . . . it was no secret within those who needed to know, within the administration. The Joint Chiefs were for it, the, ah . . . Secretary Rogers, in the meeting when it was approved, ah . . . and as far as its being neutral territory, let me say that enemy-occupied territory is not neutral. Normandy was not neutral in World War II. Do you think it was?

FROST: No, but the point was that it was—although some of the leading supporters of the war knew about it and so on—I mean, it did involve the falsification of reports and all of that, in order to keep it a secret. Didn't it?

NIXON: Oh, now, as far as the falsification of reports. That came out much later . . . the charge that the Pentagon, Secretary Laird, gave two sets of orders, ah . . . to the plane crews, ah . . . when the bombings were to take place. Ah, but he did that in order to protect the security of the operation. And, ah . . . in war, there are times when you have to, in dealing with the enemy in order even to mislead them, you may not be able to level with your friends or even with your own people. Ah, getting back to Normandy, as you will recall, ah . . . there were, ah . . . at least four different areas that the Germans expected an attack. And, ah . . . Eisenhower and his commanders, ah . . . of course, had all kinds of stories put out deliberately, ah . . . which were false, in terms of where we might attack and so forth. And then they went into Normandy. Ah, in this case, the bombing had to be secret for the reasons we did. And the other point that we have, that I have, that I have in mind here is that, ah . . . I would have preferred to have gone on television and said, "We're, now we're gonna start bombing the sanctuaries in Cambodia." But at that time and, as I said, with Sihanouk's position being what it was, I did not feel that it would be wise to do so. You're correct in saying that Sihanouk, ah . . . later in the year, in July, and when Senator Mansfield visited him in September, did not object to the bombing then. Ah, after it'd been announced.

FROST: Now, as we come towards the peace accords and so on, the offers you made in '71 and then the breakthrough in '72, ah, when the refined offer was finally, ah . . . or first agreed in Paris in October of '72, ah . . . with the concessions that we made and the concessions that that they made and so on. Leaving aside for a moment their side of the bargain at all, and leaving aside what they would say yes or no to; given the things in that offer, like the ability for them to resupply their forces, et cetera, and the sixty-day withdrawal and so on. Looking back on it now, do you think if the offer that was made, say, in August–September '72 had been made three years earlier, the war might have been shortened by three years? If you'd made that generous offer three years earlier?

NIXON: No way. Ah, no way, and I can base that on the secret negotiations and also the public negotiations that we had in Paris and the hard line that the North Vietnamese took throughout that period. I—

FROST: But if the hard line—

NIXON: I should point out—

FROST: But the hard line wouldn't have been softened if you'd made that offer?

NIXON: But the . . . but the point that I make is, the hard line that they took was that they would not separate military settlement. In other words, withdrawal of forces and exchange of prisoners of war and an internationally supervised cease-fire, from political items, they imposed as a condition in the talks they'd had with President Johnson's representatives, that any settlement had to be based on the U.S., not only withdrawal but was we withdrove . . . withdrew, throwing their and his government out of office. And that we could not agree to. And it was only when, ah . . . they finally agreed to separate the military and political issues, and this occurred, of course, in October, ah . . . of 1972, that we then had the breakthrough.

FROST: When was the moment that they really did separate? Was it a month or two before October? When did they agree to separate the two?

NIXON: They began to separate the two in September, about a month before. I think the forces that caused it were several. One, the fact that our bombing and mining—the decision of May 8—had had a military effect in closing off their routes of supply and sources of supply. Ah, second, of course, the visits to Peking and Moscow had had a subtle effect, if not a more direct effect. Whether it was direct or subtle, or both, ah . . . only time will tell. Ah, third, the fact that the South Vietnamese had fought so well during the May 8 offensive. That at the conclusion of that offensive, despite the battle surging back and forth throughout the country, that the South Vietnamese still held all the principal capitals. And finally, there was another new factor, I think, that had a considerable effect. By September, or the middle of September, when they dropped, for the first time, their insistence that we dump Thieu and his government, basically. Ah, Le Duc Tho did this in a private meeting with Kissinger in Paris. It was clear then, ah, that I had an enormous lead in the polls, and it was very likely that I would be elected in November. Kissinger played that hard. Ah . . . Le Duc Tho. He said, "Look, settle now because you can get better terms now than you may be able to get after an election. And if you wait until after an election, we will have no incentive, ah . . . to make a better deal than we can make now." Ah, that had some effect on Le Duc Tho.

But in any event, I think it was the fact that all of these forces came together at one time, finally got the North Vietnamese to reach the conclusion that they should separate the military from the political issues. Not insist on the overthrow of the South Vietnamese government, which we could never agree to. Although we did insist and we did agree to supervised elections with the Communists participating in the elections, internationally supervised. But when they finally agreed to settle the military issues and let the South Vietnamese determine their own future, ah . . . among themselves, internationally supervised, ah . . . that breakthrough having occurred led to the October 8 basic interim agreement, ah . . . and then, the events which followed thereon.

FROST: Now, the interim agreement Dr. Kissinger took with him to Saigon for Saigon's approval, and by all accounts, the result was a minor eruption or a minor earthquake. I mean, there was violent response to it.

NIXON: Well, I would say that's typical British understatement when you say a minor eruption. Ah, I'll never forget when

Dr. Kissinger got back from Paris after that October 8 meeting. Ah . . . we were over in the EOB meeting with two or three members of the staff that we have. My private office over there. Kissinger came in late in the day, and he said, "Well, the president now has three out of three." Because we'd always talked about, we wanted China, we wanted Russia, and we wanted Vietnam. Ah, and we wanted them, of course, that year before the election. He was totally confident that, based on what the South Vietnamese representatives in Paris had said, Thieu would agree to, that he would not have significant problems with them when he went to Saigon in order to get his approval, which had to be received at the highest level.

However, after he got to Saigon and presented the agreement to Thieu, Thieu raised some objections. Now, the objections, Dr. Kissinger agreed when I talked to him later, and in his messages back to me, ah . . . the objection, some of them, ah . . . made, ah . . . certainly good sense from Thieu's standpoint and were reasonable insofar as the North Vietnamese were concerned. For example, ah . . . they included the fact that the cease-fire should go into effect and be supervised immediately so that the North Vietnamese wouldn't grab up a lot of territory prior to . . . after it was announced and, ah, before it was implemented. Ah . . . it was important that the DMZ, for example, the provisions there, be more clearly spelled out. Ah . . . ah . . . the agreement had to be improved in that way. And it was particularly important that, ah . . . in Thieu's view, that the agreement had to be absolutely clear with regard to the election machinery so that there was no suggestion of a coalition government.

So after going round and round with Thieu in Saigon, Kissinger, of course, was then supposed to go round the horn to Hanoi, and, ah, he cabled back, and, ah . . . the question was, should he go on to Hanoi? And of course, my response, which he agreed with, was that he couldn't go to Hanoi unless he had Thieu on board, because otherwise Hanoi, ah, would be in a position of saying simply "We agree," and then we could put Thieu over the side, which is exactly what they wanted, and we would have had an agreement, ah . . . which would have not given the South Vietnamese a chance to survive without having the North come in and impose a Communist government on 'em against their will. And—

FROST: But what was—

NIXON: And so he knocked off the Hanoi leg of the trip—came back to Washington.

South Vietnamese President Nguyen Van Thieu has often been portrayed as stubborn, uninspiring, and inadequate to the task. Yet it may well have been patriotism rather than obstructionism that best described him. President Thieu knew a losing deal when he saw one. No one had to tell him the meaning of disguised surrender. In the end, Nixon secured Thieu's signature on the accord only by threatening to go it alone if Thieu stood silently by.

FROST: And, what form did the eruption take? I mean, did, ah, President Thieu rant at, ah, Henry Kissinger or what?

NIXON: President Thieu is not a ranting kind of a man. Ah, he's very self-controlled. Ah, Ky, for example is a flamboyant flyer, you know, with all the attributes of the . . . of the . . . well, the air force won't appreciate this, but, ah, you know the flyboys are kind of a swinging lot. Ah, but Thieu, on the other hand, was very reserved, ah . . . soft-spoken, but very firm. And what happened was that Kissinger presented the agreement to him; Thieu said he wanted to think about it; Thieu came back the next day with the changes that he wanted in the agreement. Some of them we couldn't possibly get. For example, his insistence that the agreement had to provide that all North Vietnamese forces would withdraw from South Vietnam, because the North Vietnamese wouldn't even admit they had any forces in South Vietnam. They just claimed they were supporting, of course, the VC down in South Vietnam. But be that as it may, he presented this and presented it so hardline that there was no way for Kissinger to break the bottleneck, and he felt that he would have to go back for one more meeting in Paris, in order to conclude . . . to conclude the agreement.

I should point out that when he was there, Kissinger cabled Thieu's attitude to me. I sent a very tough cable, ah . . . with a little velvet glove around the fist of Thieu. I pointed out that we considered the agreement to be a good one, ah . . . that it met the conditions I'd laid down in my May 8 speech with regard to the fact that, ah . . . ah . . . we would withdraw all of our forces in sixty days if we got our POWs; if there was an internationally agreed cease-fire; internationally supervised elections; et cetera. But in any event, I indicated to him that if he didn't go along, ah, we would have great problems in getting continued American support, and particularly within the Congress, for aid to Vietnam. That, however, was not enough at that point, but that was the first of a series of messages that became increasingly tougher and tougher, that I sent to Thieu. At the same time, I was sending very tough messages, ah . . . and Kissinger was delivering them, of course, ah . . . to Hanoi through Le Duc Tho, or directly to them. It was that kind of a way that we were—

FROST: But the initial problems in actually getting a signature on the dotted line following the draft agreement that you'd approved was, in fact, the changes that we had to request as a result of the changes requested by President Thieu?

NIXON: I should point out that President Thieu made some requests for changes that we considered to be reasonable. We told him so, and we, however, did present them to the North Vietnamese, but we did not insist upon them. He made some requests that we did consider to be reasonable. After all, he knew the North Vietnamese. He knew them better than we. He was more suspicious of them than we were, although we didn't have too much confidence in them, either, and that, of course, proved to be a very sound position after they broke the peace agreement within two days after it was signed on January 23, by their actions in Laos.

Ah, but nevertheless, as far as why the October 8 agreement, which was tentative, between the United States and the North Vietnamese, why the . . . it did not go forward . . . and, ah . . . was . . . and why it was not completed before the October 31 deadline, which we had all hoped to meet, which the North Vietnamese particularly wanted, you can't put that responsibility, as some have, solely on Thieu. He was partially responsible. But on the other hand, we agreed with some of the points that he had raised. We thought they were reasonable, and the North Vietnamese, we think, ah, at that point, and certainly after the election, became the more hard-line and the more unreachable.

FROST: But initially, we sought certain changes of points where you thought President Thieu had a point?

NIXON: Yes, that was the first thing.

FROST: And then, in fact, after the election, was that still the sticking point in fact? The changes we'd requested?

NIXON: Yes, the changes that were requested before the election became a sticking point, but what happened after the election was that then the North Vietnamese, ah . . . who probably sensed that they might be able to make a case, that it was Thieu who was blocking the road to peace, ah, that he was sabotaging the agree— . . . peace agreement that everybody wanted. They then, not only simply took a totally stonewall position with regard to even the most minor suggestions for improving the agreement or qualifying the agreement that we wanted, that Thieu wanted, but they backed off their October 8 proposal and began to raise the ante.

FROST: How did . . . how did they change the terms? This is now November, now.

NIXON: Yes, well, first, as far as the DMZ was concerned, ah . . . this sounds like simply semantics. But on the other hand, recognizing the DMZ is absolutely essential, because crossing the DMZ is basically a violation of the peace accords of years before. And they practically wanted to erase the DMZ as far as this agreement was concerned. The second point was that with regard to POWs. We wanted an absolute commitment that our POWs would be released if we agreed to withdraw our forces. They then began to hint and imply and then insist that the matter of the release of our prisoners of war, military prisoners of war, should be tied to and conditioned on the South Vietnamese government's release of civilian prisoners held in the South, which was a different issue, and which we had . . . they had agreed in October, ah . . . on October 8, could be handled and should be handled separately and on a different kind of a timetable. And then, in addition, ah . . . they, ah . . . took a position with regard to the cease-fire that would have made the international supervision and the implementation of the cease-fire a total sham. I'm reporting now the words that Henry Kissinger used in . . . when he came back from Paris after his long and tortuous meetings in November and December with Le Duc Tho and said, "I can't get an agreement." Now we have a situation where the South Vietnamese have refused to go along, as they had refused to go along before, ah, the election. Ah, now the North Vietnamese have raised the ante, where we couldn't possibly go along, even if we wanted to dump the South Vietnamese, which we didn't want to do.

One of our biggest tasks with Nixon was to think through what would have happened in Vietnam had he remained in office. This historical jigsaw puzzle may never be resolved because not all the missing pieces will ever be found. Would Congress have passed a War Powers Act with a strong president in the White House? Could Nixon or some hard-line successor have gone back into Vietnam to prevent corruption of the deal? Nixon offered his judgment. I played devil's advocate.

FROST: And so then, in your secret letter to President Thieu of November the 14, where you promised that, if Hanoi failed to abide by the terms of this agreement, "it's my intention to take swift and severe retaliatory action." What did you have in mind at that point?

NIXON: Well, that was simply a letter that had as its purpose giving him the self-confidence that he needed to sign the agreement. He didn't trust Hanoi. He thought they would break it. And he had good reason to not trust them, as I've indicated. Ah, I therefore, wanted him to know, ah . . . that he had my commitment, my personal commitment, ah, to take whatever action I could as president, having in mind, of course, the reservation that he knew that I always had to have in mind . . . I'd have to have congressional support on the appropriations side. But in any event, I gave him that assurance, hoping that that would get him to be more reasonable in terms of accepting the terms of an agreement.

FROST: Did you think you'd get that approval if it ever came to it?

NIXON: Oh, yes, ah, I felt, ah . . . I felt that if we got a good agreement, a reasonable agreement which got back our prisoners of war, ah . . . which provided for a cease-fire on both sides and agreement that everybody adhered to . . . ah . . . I thought, ah . . . that we would be able to get congressional support, ah . . . for continuing to provide economic assistance and military assistance to South Vietnam, which the agreement provides for. You know, the agreement provided as far as reinforcements were concerned, into the South, there could be none. It did provide, however, that as far as North Vietnam and South Vietnam were concerned, that each . . . that material could be replaced on a one-on-one basis.

FROST: But you feel that you could get congressional approval for "swift and severe retaliatory action" with full force?

NIXON: I, ah . . . with regard to the swift and severe retaliatory action, I feel that if the North Vietnamese, which they had so often done in the past, flagrantly and blatantly violated the agreement, that I could go to the country and to the Congress and get the support that was necessary to bring them into line.

FROST: In other words, you'd have—

NIXON: I didn't think . . . I didn't think, in other words, to use the Chinese term that, ah . . . Chou En-lai often used with me, I didn't believe this was an empty cannon. It wasn't an assurance that I didn't feel that I could keep.

FROST: But you felt—

NIXON: And I was . . . and I'm making it clear to Thieu that it was my intention to do this and that I would ask for it, and I felt that I could get it.

FROST: You felt you could go to the Congress and, ah . . . address the nation and say, "The war was over, we had peace with honor, we must go back in again and do this"?

NIXON: Well—

FROST: That's "swift and severe retaliatory action." And that the country would have gone with you?

NIXON: Let, let me, let us understand what we're talking about here. I wouldn't have had to go back . . . go to the Congress for the purpose of getting approval of the action to take. As president I, ah, could have taken that action at that time before the War Powers, ah . . . Bill was passed . . . Resolution was passed in October of 1973. I could have taken that action, ah . . . and, ah . . . then the only question is whether or not the Congress would have, ah, aborted the action by ending funds for it, which they could have done. However, I felt that if the North Vietnamese broke the peace agreement in an open way, and if the South Vietnamese complied with us, as we insisted that they do, ah . . . ah . . . right on down the line without division, I was confident that I could take action and that I would get support for keeping a peace that we had won.

FROST: But the difference . . . the difficulty of you going on television at that point would have been . . . all the other times you went on television, although you were announcing the military ventures, they were all part of getting out. At that point, we would have been out and you would have been going on television for support to go, maybe only with airplanes and bombing, but to go back in again. Do you think you could have—

NIXON: —only air strikes. As far as the actions of North Vietnam against South Vietnam were concerned, the balance of power between the two, the South Vietnamese have demonstrated, ah, that they could hold their own and they continued to demonstrate it for over two years after the peace agreement. That they could hold their own against the North without any support from us, insofar as airpower or other things were concerned. Now, I felt that the use of airpower, ah . . . ah . . . in the end there were massive violations on the part of the North Vietnamese, would have had public support. Because, the people having supported the action we had taken previously to get the peace, I think, would have supported what we had to do to keep the peace. It would have been a hard case to make, ah . . . and I told Thieu this. That it would, of course, would be . . . I told him it would be a hard case to make to get the support for him economically and militarily unless he signed the agreement. He always knew that it was a hard case to make, ah . . . for a president to get support from the country for military action and military aid. But I was convinced that in making, ah, in telling Thieu, as I did in that letter, and in others as well, that I would ask for and I would take retaliatory action that I would have public support for doing so. I would have broke the case strongly; it would have been swift; it would have been massive; and it would have been effective.

FROST: And during the delay that went on during that particular period, ah . . . before the signing in January, massive amounts of arms were shipped to President Thieu to strengthen the South Vietnamese position and so on. Was that one of his conditions for going along, in fact?

NIXON: As a matter of fact, ah . . . the shipment of arms to President Thieu in that period was, ah, ah . . . not in any way, ah . . . a violation of, ah, any congressional mandate or . . . inconsistent with—

FROST: Oh, no, no. I'm merely saying—

NIXON: —disagreement, because you see, the peace agreement had not—

FROST: I'm merely saying—

NIXON: —had not yet been entered into. Ah, as far as our shipment of arms was concerned, ah . . . we knew that the North, based on what they had displayed on that May 8 offensive, had a lot of new sophisticated weapons in their inventory that they'd received from the Soviet Union. The purpose of our shipment of arms to the South Vietnamese, since we were getting out, was to be sure that once the cease-fire was agreed to, that the South would have at least an equal chance against the North. And so we simply provided for Thieu and the South Vietnamese what we hoped would be at least as good equipment—arms, ammunition, et cetera, et cetera—as the North had. And, ah . . . I must say, I think that Thieu's attitude in getting them eventually, while he did so reluctantly, even to the last, to go along with the peace agreement.

FROST: But did he set it as a condition that "I must have more . . ."?

NIXON: Oh, no, no, no. No, he didn't set it as a condition.

FROST: But now—

NIXON: He suggested it, but not as a condition. I wouldn't have accepted that.

Nothing about the Christmas bombing seemed to make sense at the time Nixon conducted it, nor even five years later. The North Vietnamese had already indicated their agreement with the deal ending the war that had been presented in October. American POWs would soon be released. An in-place cease-fire would shortly begin. Perversely, the bombing seemed intended to quiet President Thieu's fears rather than to achieve concrete results. Yet Nixon's whole rationale for his approach in Vietnam rested upon his ability to resume military operations should the North Vietnamese defy the treaty's terms. I felt close questioning of him was in order.

FROST: Was, in fact, one of the motives of the Christmas bombing, we've been through the sequence of events before, but in addition to what you had to say, by that Christmas bombing to North Vietnam, was one of the side motives as well to show President Thieu the strength of American support that he had? That one of the motives, or one of the side effects of this, was, in fact, to reassure President Thieu as much as attack North Vietnam?

NIXON: Well, before we answer that . . . that, ah, directly, which I, of course, will, ah, let's, while it seems like just a semantic point, let's understand what the Christmas bombing was. There was no bombing on Christmas, of course. We had a Christmas truce for forty-eight hours. At the end of that truce, I ordered the biggest strike of the war. And within one day after that strike, ah, on the twenty-eighth of December, ah . . . it worked. The North Vietnamese agreed to come back and negotiate without conditions, and that meant going back to the October 8 proposition and in fact we were able to improve upon it.

Now, with regard to the effect of the Christmas bombing. There isn't any question, ah, that it had, I believe, a salutary effect on Thieu, in that it indicated that the United States, ah, was going to still take action against the North Vietnamese, if the North Vietnamese backed away from what we considered to be not a formal agreement but at least, ah . . . tentative agreement, which they had made . . . many, many months before. So it helped his morale. Let's put it that way. However, I think what we have to understand about the difficult bombing decision was that when Dr. Kissinger came back on December the sixteenth—we started the bombing on the eighteenth—ah, it was one of the most depressing days of all the White House years. And, ah, because he had worked so hard, he had worked day and night. The stamina of the man is unbelievable. That's when he can outlast anybody. Ah, when it comes to negotiations, as the Chinese, the North Vietnamese, and the Russians have learned, as well as many others, who have stamina. But in any event, after he came back, he was terribly tired, and he indicated that we had no choice, in his opinion, in view of the North Vietnamese having raised the ante and now putting forth not their October 8 agreement but one which didn't provide what we insisted upon—the unconditional release of our POWs—which would erase the DMZ, which did not provide for a cease-fire in place, because it made it a sham because of lack of adequate supervision. Ah, and which was not as clear as it should be with regard to Laos and Cambodia, which later cam— . . . became issues. Ah, he decided that we had to do something. The only disagreement that we had at that point—I say "disagreement"—discussion was whether the bombing should be preceded by a television speech announcing it—

FROST: And he thought it should, didn't he?

NIXON: Yes. He recommended a television speech at first. And then I had a long talk with him, several long talks with him, about it. He said, "You must rally the country again." And I said, "Well, what do we rally them about?" Ah, and well, he said, "Well, you rallied them on November the third, 1969, and you rallied them when you went into Cambodia, and you rallied them when you went in May 8, now rally them again." I said, "The difficulty now, Henry, is that after the American people had been told"—ah, as they had on October 26—"that peace was at hand." A statement which, incidentally, was totally true, and he made it. But a statement which, ah . . . he . . . who is hard on other people in his criticism, as I am hard on him, but is also hard on himself, as I am also . . . try to be hard on myself, that he realized boxed us a bit into a corner. Because by saying "peace is at hand," it put the North Vietnamese in a position where they realized that, ah, we had to have peace. That we had to negotiate. We had no choice, no option. And, ah . . . it put the American people . . . it gave them a euphoric view that, ah . . . another meeting or so and it was over. As far as the American people were concerned, after the election, after the statement was made, after we had reduced the bombing and so forth, ah . . . they wanted to put the war behind them. And now to go on television and to tell them, "No, we were wrong, the North Vietnamese reneged in their words and now we are going to force them back to the bargaining table," I said, "would have a devastating effect on American public opinion and would have the wrong effect on the North Vietnamese."

And this is the point I wish to make particularly. The main argument, which to Henry Kissinger, at that point, was absolutely conclusive and left everything else to be irrelevant . . . the main reason you couldn't go on and make a public statement with regard to the bombing was that in making that statement, it would have to be said, "We are doing this because they have not agreed to settle on certain terms that they agreed to earlier and that we believe are reasonable now." That would have put them on a public spot of having to cave under military pressure and very difficult for them to do so. Also, it would have put the Chinese on a spot. It would have put the Russians on a spot, if we had . . . if I had made a public statement, they would have had to have public statements and, who knows, even demonstrations in support of the . . . North Vietnamese.

And then finally, as far as the American people were concerned, we would have . . . might well have run into a situation by making a speech about it, that American support for the action could not be garnered. What would only work at that time was success. What could only work was the action itself. I felt that the American people, ah . . . ah . . . without any question did not want a peace that was wrong. That basically was a bugout, that was a cave-in and a surrender to the North Vietnamese and fighting the war solely to get back our prisoners of war, which someone suggested was the deal that we should make, and one which, incidentally, the North Vietnamese would never accept, even when it was presented to them.

FROST: When was it presented to them?

NIXON: It was presented to them in a very curious way. Not very seriously. This was back in 1971, when we were making proposals early in '71 and through that period, ah . . . which were very reasonable, with regard to withdrawal of our forces. By May of 1971 we said we would withdraw our forces and we did not insist on their withdrawing theirs at the same time, provided we had an internationally supervised cease-fire, et cetera, and a return of our prisoners of war. Ah, in, ah, October, no, in May and July, Kissinger said we'd do it in six months. We set a deadline. This on the secret channel. In October, again using the secret channel, I sent a message directly to the North Vietnamese saying we would in six months, all of this in the '71 period.

Senator McGovern, who was then very actively seeking the Democratic nomination, of course, which he later got, took a trip to Paris and came back and said that he had gathered the impression that if the United Sates would agree to withdraw its forces, that we could get back our prisoners of war. That we could make that kind of a deal. Well, the next time Kissinger met with them, and I talked to him about it, I said, "Why don't you put that to 'em? Let's see." I said, "I don't believe that, first, they'll take it." And Henry totally agreed that they would not accept it, for reasons that you'll see. But I said, "Let's find out whether or not this is something that is serious." He presented it to Xuan Tri. Xuan Tri totally stonewalled it. And then Kissinger said, "Well, Senator McGovern said that you will agree to, ah . . . returning our POWs if we will withdraw." And Xuan Tri coldly answered, according to Henry's report, "Well, what Senator McGovern says is Senator McGovern's problem." Because, you see, throughout '71 until we finally broke through in October or September, I should say, of 1972, the North Vietnamese would not agree to separate the military issue, like which withdrawal of our forces and prisoners of war amounted, from political issues. They insisted, always, that a condition for any settlement was that we must dump Thieu and allow the imposition of a coalition government.

The notion that early in his administration, Nixon might have been willing to swap an American withdrawal for the return of POWs took us all by surprise. Yet the North Vietnamese refusal to entertain the idea indicates that so long as the Thieu government was in power in South Vietnam, the war would continue. But it also indicates Nixon's willingness to sacrifice U.S. credibility, not to mention a bunch of dominos, for very little.

FROST: When we mentioned earlier the "swift and severe retaliatory action" and so on, um, that you mentioned in your letter to President Thieu, you mentioned just then that in April '75 there was the complete breakthrough by the North Vietnamese. Had you been in the White House in May 1975, do you think you could have done what you said you would have done in those other circumstances, a major North Vietnamese violation, do you think you could have done something to save South Vietnam at that point in May 1975?

This question drew the first of several responses by Nixon blaming the Democratic congresses elected in 1972 and 1974 for the fall of South Vietnam. It is a case that seemed frivolous at the time but that has grown in credibility as the North Vietnamese themselves have borne witness to their having kept an eye on the U.S. Congress and the enfeebled presidency before launching their 1975 climactic offensive.

NIXON: No. No, because you see, the Congress, the Congress had very effectively, ah . . . taken action which limited the president's power, ah, in the area of defending the interests of the United States in making with military action where those interests are threatened. Ah, and I refer to two cases here: one, an August 15 cut off of all authorization of any bombing in any part of southeast Asia. That was in 1973. And then, second, ah, the War Powers Resolution, which, ah, they passed in, ah, October of 1973, ah . . . which in effect says, ah . . . declares that when the president does initiate any action, military action to protect the interests of the United States as—which he has a constitutional right and power to do—that he cannot continue that action beyond sixty days unless the Congress specifically, by positive, votes for continuation of the action. So, under the circumstances, President Ford's hands were tied. He could not bomb . . . and if he were to go back in on some other basis, the War Powers Resolution was hanging over him. It would have been a very difficult situation for him.

On the other hand, let me point out that when you say, "Could it have been saved?" and the answer is "It certainly could have been saved," and you can't blame President Ford for this . . . the blame, ah, has to be placed where it belongs, and it belongs with the Congress of the United States, not simply because they had done what they'd done in 1973 with regard to limiting or cutting all right to bomb, ah, order any bombing in Southeast Asia, not only because what they'd done with regard to the War Powers Act, but because they cut off Thieu's water, in effect, to speak in the vernacular, by refusing to comply with the peace agreement, our obligation to replace his equipment on a one-on-one basis, so that when the North Vietnamese started their offensive, as Lee Tuan pointed out, Thieu was fighting basically a poor man's war, with only sixty percent effectiveness that he'd had previously and in many areas in those last final weeks. From the reports that I have read you there isn't any question but that the North Vietnamese had a two- to three-to-one advantage in heavy guns, heavy artillery, tanks, and sophisticated weapons over the South Vietnamese.

And why did this happen? It happened because the Congress refused to grant President Ford's request to provide the funds for the South Vietnamese to defend themselves. If the South Vietnamese had had the necessary equipment, I believe they could have held on. I say this despite the fact that, ah . . . as we have discussed in other periods, that Thieu had political problems, ah, that the South Vietnamese army . . . was spotty insofar as its efficiency in certain areas, ah . . . but they'd held out for over two years, and they could have held out longer, ah, but they couldn't hold up against a massive invasion in total violation of the peace agreements from the North.

FROST: But even in January of '73, just before the accords were signed, there was obviously doubt somewhere within the administration about their ability to survive, because, on the one hand, Secretary of Defense Melvin Laird said (this is in testimony to Congress), "I can say to you completely and totally that the South Vietnamese have a better military capability than the North Vietnamese." He went on, then he said, "They're fully capable of providing their own security against the North Vietnamese." And then said, "Vietnamization makes possible the complete termination of American involvement." Involvement, not presence on the ground but involvement in the war. And at the same time in a letter on January 5 to President Thieu, you said, "We'll respond with full force should the settlement be violated by North Vietnam." Now, on the one hand, Melvin Laird seems to be saying that the South Vietnamese could, could fully defend themselves come hell or high water and there'd be no American involvement, and, on the other hand, you were saying, really, in your letter to Thieu that American assistance would be needed and provided. I mean—

NIXON: Ah, first I didn't say that it would be needed. I said that if in the event that the North Vietnamese broke the peace accord, in a blatant and flagrant way, that the United States would be in a position where we would take strong action. And let me also point out . . . point out . . . but . . . but . . . let me point—

FROST: But why would we respond with full force if it wasn't needed, though?

NIXON: Well, I . . . I think we're just quibbling about words here. The point was that I was hopeful at that point that it would not be needed. Ah . . . we did not expect that, ah . . . we would be faced with a situation where the North Vietnamese would violate the agreement with, ah . . . so much power that the South Vietnamese would not be able to defend themselves. And we proved to be right for a great period of time. The South Vietnamese, ah, did as I pointed out. They held their own through the balance of 1973 and through all of 1974 until April of 1975 without any American involvement. You used the word *involvement.* But Secretary Laird did not mean, of course, to exclude American military aid in terms of equipment—

FROST: No, but I think he—

NIXON: —he was very strongly for that—

FROST: —the context of his speech clearly excludes what we've discussed that was in your mind, which was bombing. And his speech clearly, on January the eighth, excludes the idea of bombing . . . which would have to—

NIXON: — but he was basing that, he was basing that on the assumption that the peace agreement would be kept.

FROST: —he was saying that come hell or high water—

NIXON: —that's what he was basing it on—

FROST: —I mean that . . . he was saying . . . but—

NIXON: Whatever, whatever . . . whatever he has said, ah . . . the record will speak for itself, ah . . . I will simply state my own view, that had . . . had the United States Congress, ah . . . ah . . . clearly apart from the War Powers Resolution, where they perhaps had other reasons to vote as they did, clearly apart from the cutting off of the authority to bomb in Southeast Asia, which they cut off, but on the key point of providing for a friend, an ally, they . . . at least an equal chance to defend itself, for the Congress of the United States to refuse to do that when the president of the United States requested them to do so in the latter part of 1974, that led to and triggered the total collapse of South Vietnam. Ah, and I wouldn't, don't blame the South Vietnamese a bit under these circumstances. They can be terribly brave, but if, when you have a two- to three-to-one advantage looking at you in terms of tanks and guns, you would run, too.

Many scholars dispute this analysis. The claimed North Vietnamese advantage in equipment is regarded with skepticism by those pointing to North Vietnam's own complaints about the failure of its Chinese and Soviet allies to deliver needed weaponry. Others point to the complete collapse of South Vietnamese resistance after the initial North Vietnamese advance in the central highlands. There may have been token resistance in a few areas, but eyewitness accounts from the period also note the South Vietnamese units that capitulated to the North Vietnamese, who were driving through their areas in passenger vehicles and proclaiming their victory through bullhorns.

FROST: So that in fact you felt that American assistance, military assistance, would have to be permanent if South Vietnam was to survive.

NIXON: Well, permanent as long as the Soviet Union was providing permanent assistance to the North, yes. And that's got to be true in other parts of the world as well. I think that we have to understand, when we talk about détente, for example, is that, ah, it's a two-way street. You can't say to the South Vietnamese, ah, ah . . . that we're going to have you take care of your own needs, develop your own economy to the point where you can pay for everything yourself, and we will not help you, ah . . . as against any of your neighbors when one of their neighbors, in other words, the North Vietnamese, in this instance, are being massively helped by another, by a major power, the Soviet Union. To a less extent by the PRC. In other words, when you talk about whether or not the United States should or did take on a commitment to help South Vietnam forever, the answer is, of course, not forever, it would depend on the circumstances, but I would certainly say that the United States did take on an obligation, and should have taken on the obligation, of helping the South Vietnamese to the same extent that the Soviet Union was helping the North Vietnamese and to provide them the means to defend themselves. Just as we're doing, for example, for Israel and other countries.

The problem with the domino theory was that, by the time our interviews were conducted, it was evident that the loss of Vietnam had had few if any consequences apart from Laos and Cambodia. The dominos—Singapore, Malaysia, Indonesia, the Philippines, and others—proved stronger than many people had suspected. Extending the domino specter all the way to South America seemed positively loopy.

FROST: You've talked about peace with honor. You've talked about the reasonable chance to survive and so on, and that's what certainly your supporters would say was achieved in Vietnam. The cost we've said, was probably, during your administration, seventy-five billion dollars, 15,000 American lives, 138,000 South Vietnamese, half a million Cambodians, 590,000 North Vietnamese, and so on. As we look at that cost, and as we take into account the body of opinion that feels that that was not so much a peace settlement as a piece of paper, of which tragically nothing remains today, do you still feel the cost was worth it?

In many people's view, Nixon's Vietnam policies were not only not worth it but had become positively dysfunctional. What was important to Nixon was détente with the Soviets and the beginning of a relationship with the Chinese. Both could be endangered by developments in Vietnam. And both countries knew that Nixon wanted out, requiring only a transparent agreement to end the fighting. But Nixon remained insistent.

NIXON: It was worth it in terms of the period in which I had the responsibility. I mean, one can only judge his own actions, why he took such actions, whether he felt that those actions, once taken, would stand historical scrutiny. All wars are terrible; if there was ever one in which a country had motives that were decent, motives that were unselfish, this was that kind of a war. We didn't want anything in Vietnam. We didn't want any bases or anything of that sort. Ah . . . we weren't asking for domination of the country. We don't want any of their economic resources. It's a relatively poor country, for that matter. Ah . . . but we were trying to keep our treaty commitments, and we were trying, of course, to deter what we thought was aggressive action, which, if not deterred at that point, might gather steam and flow over the other part of the . . . of the Southeast Asian area, now the—

FROST: Did that—

NIXON: —point is the fact that some of that did not happen as soon as we think, ah . . . I suppose makes the case for why we shouldn't have done it in the first instance. My point is perhaps we will be sitting maybe two years from now, who knows? Here or someplace else. Ask the question then. My point is that as far as the Untied States was concerned, the war in Vietnam was a terribly difficult war, I am sure, for President Kennedy to make the decision initially to commit combat forces to. It was a terribly difficult war for President Johnson, to continue on a massive basis. It drove him out of office. It was not an easy time for me, because I hated every minute of it and would have done anything I possibly could to bring it to an end so I would not have to . . . so that we could concentrate on more important things. Ah . . . like, for example, our opening to China in the field of foreign policy.

But my point is that as far as the war was concerned, at the time that I made the decision, it was the only, in my view, right decision to be made. Ah, it, ah . . . it came up, for example, in my talk with Chou En-lai. He took what I expected him to do. He said, "We note some press reports to the effect that you're coming to China for the purpose of targeting us to help you on the war with Vietnam. We want to make it very clear that our position is one of support for our comrades in North Vietnam and in, ah . . . South Vietnam, the so-called Viet Cong. And we believe that you should get out." And, ah, my response to him was that I respected his opinion, and in our communiqué, which was very unusual, they set forth their position, what was that, and we set forth our position, rather than weasel-wording it and trying to camouflage the differences or paper them over. But I said to him, I said, "If the United States doesn't prove to be an ally or friend one can trust in Vietnam, we aren't an ally or friend that other nations may be able to trust." I said, "That's what's on the line there, and therefore, we're going to see this through to an honorable conclusion."

FROST: But we're sitting here after two years and the domino theory hasn't happened in that part of the world yet, and you said maybe we should talk about it in another two years. If we talk about even in another five years and it hasn't happened, bearing in mind what a close call you said the decision on Vietnam was, the cost in lives, the cost in divisiveness here, and all of that. If the domino theory hasn't happened in five years, would you then feel that you had to say, "Well, with the hindsight of history," which no president has, but "With the hindsight of history, our policy was wrong"?

NIXON: Well, naturally, I can see that you are trying to press me into the position of saying, and it's perfectly all right to do it, because it's part of the present-day thought about this terribly difficult war, and I recognize, in all candor, that, ah . . . I could take . . . I could state a much more popular position. Let's face it. Let me be quite candid about it. Ah, the most popular position to take on Vietnam, if I was simply playing to the votes and playing to the popular opinion in the world, was to bug out and blame it on Johnson and Kennedy. And I didn't do it. Now, the most popular position for me to take now is to say the whole venture in Vietnam, everything that we did, was a waste of men, that it . . . it . . . showed the United States at its worst. It cost us a great deal of money. We were morally wrong to be there . . . go there in the first place. Morally wrong to continue it as long as we did, and it wasn't worth it. And I could say that. And many, perhaps, of those, and it's probably a majority of our viewers who agree with that, who applaud some of my critics, but I'm not going to say it. I'm not going to say it because I don't believe it.

I also don't believe that this was a war that . . . I can put up the advantages and disadvantages and say overwhelmingly, this is a war that had to be fought and that we had a successful outcome. I can't say that because it was a very complex situation. It was complex at the beginning. It was difficult throughout. I know, I know what Johnson went through and how he agonized over the war. He agonized over it more openly than I did, but, ah . . . I felt it, I think if anything, even as deeply or more deeply than he did, as we often talked about it. But I do know that, that, ah . . . the situation with regard to whether it was worth it or not, the point was that the decision that we made at that time, the decisions, certainly, that I made, I think, were the right decisions to make. I think the peace agreement was the best peace agreement we could get. It's true, it lasted for only two years. The responsibility of its not lasting is on the Congress of the United States, not on President Ford, not on those who made the decision previously, and not on the people of South Vietnam.

FROST: No, the point I was making was that there are people, exactly as you say, who would say that everything about this war was wrong. It was a tragic waste from the beginning to end. It was immoral. It had achieved nothing. It destroyed a country and so on . . . would say all of that. But, leaving aside people who feel just one hundred percent as much as that, what I was saying was, that if we went into this war, or prolonged this war because of the belief in the domino theory and in the credibility of America, and if it turns out that no domino theory so far, and no domino theory in the next few years, and America's credibility and their ability to take on Russia, her ability to talk to China has suffered no impediment, in fact, perhaps the further impediment's been removed, then we have to go with the critics, because the war was a waste, not on moral grounds but on intellectual grounds as well.

NIXON: Now, what you're talking about now is, ah . . . ah . . . what, ah . . . our former President FDR used to say "is a very iffy question." If there happens within the next five years . . . let me say, I'm greatly concerned what will happen over the next five years. Ah . . . I am not going to be critical. I'm not going to lob any suggestions in or criticisms in to those who have the awesome responsibility to make the foreign policy decisions, to President Carter and his team at the present time. Certainly not at this time. Ah . . . because, ah, I wish them well and the whole world should be wishing them well and praying that they do well, and I'm among those who have both of those sentiments. But on the other hand, I think that on the line today as a result of . . . one of the results of Vietnam, and, ah, perhaps on the line for other reasons as well, ah, which we're quite aware of, is whether the United States is going to be a responsible free world leader in the years ahead or whether we are going to turn away from leadership.

FROST: Yes, but—

NIXON: Turn inward, turn more isolationist. Turn away from our commitments. Ah, I can well understand the arguments of those who say, "Don't make any further commitments." I can well understand the arguments of those who say, "If you have a commitment, do just as little as you can to keep it."

FROST: But surely the great danger . . . this great danger of getting public support for American involvement in a worthy cause in the future, as if the past causes have been judged unworthy by the American people. In other words, you lose support for Israel if you fight the wrong war; then people probably won't support the right war, so that therefore, in that sense, you could argue that Vietnam was counterproductive and that the low figures in the Louis Harris Poll for the number of people who'd fight for Israel is due to the fact that we fought too long the wrong war.

NIXON: Well, I guess we'd have to psychoanalyze all of the thousand people that were interviewed in the poll. Ah . . . I could only respond to that by saying that the war in Vietnam was not a popular war. Ah, just as the war in Korea was not a popular war. Ah . . . as far as the war was concerned, ah . . . I think we could discuss this for the balance of the time we have on the air—

FROST: Twenty-four hours altogether—

NIXON: But as far as the war was concerned, I am not going to be in the position of saying that President Kennedy, President Johnson, President Nixon, and after that President Ford, who, of course, the war was over then, but continued to support South Vietnam from an economic and military standpoint. That all of us were engaged in the wrong course. Ah, we cannot be sure. I do know what was wrong. I do know that the Congress was wrong in having America fall down on its commitments. That the Congress has to be held responsible for that, and that had the Congress kept its commitments, the commitments we had made to South Vietnam, that we couldn't even be discussing this esoteric question of "What's going to be the situation five years from now?" as to whether the United States is going to have credibility or more or less credibility in its commitment to other countries and because of what we did in Vietnam.

FROST: But nevertheless, Congress didn't keep us in Vietnam.

NIXON: As a matter of fact, the Congress lost it. And that's the tragedy and they have to take the responsibility for it.

FROST: Don't you think it was lost the day that that accord was signed? Perhaps even the day you came into office?

NIXON: No, on the contrary. I think that the . . . I think that if the Congress had provided the funds necessary to the South Vietnamese, ah, in the period of 1974, late 1974 and early '75, that they so desperately needed, that the South Vietnamese would have survived. Ah, true, not a democracy meeting our standards or Britain's standards or the standards of most European countries, but yet, as I've said, at least a country in which there is some freedom of religion or press, et cetera, and some elections rather than none. And what they have now is a terrible disaster, and no one knows it because I see very few of those who were talking about the situation insofar as the press censorship under the Thieu regime reflecting on what was going on in South Vietnam now, because there is no free press at all there now, of course, as you know.

FROST: But let me put really in conclusion, before we move on, two different thoughts. One's a personal thought of someone who worked very closely with you, said that, ah . . . "President Nixon felt very strongly about not losing and it was that in his makeup that made him unable to recognize that the war in Vietnam was a losing battle." Do you think, as you gaze into your innermost soul, there could be any truth in that?

NIXON: No, my decision, ah, decisions with regard to Vietnam were not made . . . made on the basis of, ah . . . simply my personal ego of not wanting to quit, and no . . . I wasn't concerned about my quitting, I was concerned about America's quitting. I wasn't concerned about my credibility or prestige, or what have you, because the presidency is transitory. I certainly was concerned about America's credibility in the world. And I didn't want America to be a quitter.

FROST: If the Vietnam War had not gone on throughout your presidency, there probably would have been less, much less, domestic discord, and the unifying policies that you adopted at the beginning might not have led to an atmosphere of polarization and many of the so-called abuses of power might never have occurred or come to light or been necessary. In that sense, someone has said, I wonder if you agree, that in that sense, perhaps you were the last American casualty of the Vietnam War.

NIXON: A case could be made for that, yes. Ah, there isn't any question but that in the conduct of the war, I made, ah, enemies who were, from an ideological standpoint, ah . . . virtually, ah, well paranoiac, I guess. Oh, the major newspaper publisher told Henry Kissinger one night, right after the peace settlement, "I hate the son of a bitch's guts." And, ah . . . naturally, coming right after the time that we had been able to have the peace settlement is an indication how deep those passions ran. Ah, because that kind of attitude developed over a period of years. I mean, my political career goes back . . . over many, many years. But the actions—and many of them I took with great reluctance, but recognizing I had to do what was right—the action that I took in Vietnam, one, to try and win an honorable peace abroad, and, two, to keep the peace at home. Because keeping the peace at home and keeping support for the war was essential in order to get the enemy to negotiate. And that was, of course, not easy to do, in view of the dissent and so forth that we had. And so it could be said that I was one of the casualties or maybe the last casualty in Vietnam. If so, I'm glad I'm the last one. I hope no more.

And I must say that in expressing that hope, I do so knowing that that hope will never be fulfilled, because at the present time there are over 700,000, by the latest and most conservative estimate, people in reeducation camps in South Vietnam. Thousands have been probably starved to death. Ah . . . religious freedom, to the extent that there was any, is virtually wiped out. It's a tragic situation, which has happened, what has happened to them. And there are going to be more casualties in Vietnam. But as far as the Americans are concerned, perhaps I was one of them, too. But on the other hand, I do believe that looking back over that long and difficult period, that the decisions that I made at the time had to be made, that they were the right decisions.

FROST: And on that fact, history will have to be the judge.

11

KISSINGER

Richard Nixon and Henry Kissinger formed one of the most intriguing foreign policy teams in the history of the U.S. presidency. From Vietnam to the Soviet Union, from China to the Middle East, from Chile to Bangladesh, they played a continuing high-stakes game routed in the school of "realism." But while Americans and international observers debate the merits of détente with the Soviet Union, normalization with China, and Kissinger's shuttle diplomacy in the Middle East, others have remained more fascinated with the personal relationship between the two than by their actions in the international arena. Their own relationship could be described as "love-hate." Their grandiose international visions were often juxtaposed by a smallness of spirit vis-à-vis each other. Yet, like that of Gilbert and Sullivan, their music was never more appealing than when they acted in collaboration with each other. With their team severed by Nixon's premature withdrawal from the presidency, Kissinger began a long but steady descent in the eyes of his Republican constituents. Nixon, of course, remained an active commentator on international affairs until his death, but the heady days with Kissinger were gone forever.

FROST: What did Chairman Mao make of Henry Kissinger? He must have found him an interesting phenomenon.

NIXON: They hit it off extremely well. You could tell from the bantering tones that he was . . . whenever he referred to Henry Kissinger, ah . . . I think one of the first he said was that, ah . . . Kissinger, ah . . . was a philosopher and, ah . . . he was as doctor of philosophy, as he put it, because I had referred to Mao as a philosopher and before . . . and before I arrived and I said . . . no, he's a doctor of brains . . . then, as we began to discuss the situation in various parts of the world . . . you have to understand this . . . Mao didn't know the world . . . not well. He had studied it, but he did not know it. Chou En-lai, on the other hand, did know the world. He had traveled broadly. Mao consequently left the discussion of the various sections of the world pretty much to Chou En-lai. He talked more in philosophical terms, and then Chou En-lai would take this little phrase or that little phrase or the other one and then use that to build a policy on. Ah . . . but when the policy . . . the question of the Mideast came up, he said, "I know the Mideast is a very difficult problem." He said, "it's particularly difficult for Dr. Kissinger because he's a Jew." And I said, "Mr. Chairman, it's true that Dr. Kissinger is a Jew, but he is an American first, and you will find that, in his dealings with this problem and with any other one, he will put the interest of the United States first."

Well . . . the other point . . . oh, there were so many that were . . . that came up involving various personalities . . . but one on Henry that was, I thought, particularly interesting was this whole business on secrecy and, ah . . . ah . . . as he looked over at Henry, he said, you know he really doesn't look like a secret agent . . . And I said, well I don't know any other man in the world that could possibly have gone from Paris twelve times and gone to Peking one time secretly without anybody knowing it except possibly two pretty girls. Mao listened to the translation . . . He broke into a sig . . . smile . . . two girls in Paris . . . and then Henry stepped in . . . he said, ah . . . well . . . I was only using them for a cover, and then I said, well . . . I certainly couldn't use pretty girls for a cover in my position, and then En-lai said, "You sure couldn't do it in the new election year?"

FROST: In terms of this close relationship with Henry Kissinger . . . ah . . . you must have had disagreements, I suppose. What were the most important oncs?

NIXON: Well, I don't think it serves a useful purpose for me to try to search my memory to think of every time Henry Kissinger and I had a conversation where we didn't agree. Ah . . . when he came on board . . . one point that I emphasized to him is that I had sat in the meetings in the National Security Council for eight years under President Eisenhower, and the council, whenever he was in Washington, used to meet every week. And I said one of the things that was wrong about those meetings is that we would meet when there was nothing to meet about many times. And that the National Security Council staff would have a paper and at the end of the staff, which would be Bobby Cutler in earlier years and Dillon Anderson in later years and Gray in later years, they'd read the paper, and they were the most boring things in the world. Very seldom did we have give-and-take discussion where the president got the benefit of different views from the members of his team. So it was Kissinger's responsibility not only to give me his views when they differed from my own but also to see that the secretary of state, Bill Rogers, Mel Laird, the chairman of the Joint Chiefs. Ah . . . sometimes the CIA would present their views. We tried to do that. Kissinger was . . . became less and less, ah . . . willing to do so as time went on because of the fear of leaks. But on the other hand, I found those sessions very valuable.

Now, as far as disagreements are concerned, I should emphasize what disagreements we had were always on tactics and never on strategy. People who think that Kissinger was basically a soft-liner and I was a hard-liner, ah . . . just don't know, ah . . . what each of us believed. For example, in the soft-line/hard-line dialogue, the first crisis, so-called crisis, I guess we should say, 'cause we had so many that this gets now into perspective, ah . . . was very early in the administration, and the EC-121 was shot down by the North Koreans, and, ah . . . all the crewmen were lost. Ah . . . it was an unarmed reconnaissance plane, and it was shot down over international waters. Options were presented, ah . . . everything from doing nothing to, ah . . . taking out two or three airfields. Kissinger came down hard, this is in that very . . . in about the third month of the administration, for the hard option. He said that the Russians or possibly the North Koreans were testing us. If the Russians weren't testing us, the North Koreans and the Chinese and the Vietnamese are all going to be watching to see how we reacted to this. And he said we must react strongly, and he advocated the option of airfields as a result of this. I considered the option. Frankly, I tilted towards it. Ah . . . Bill Rogers, Mel Laird thought it was too early in the administration for us to take us an option of this sort. Our ambassador in South Korea thought it would be a mistake . . . that it would have Kim Il Sung, the North Korean, take some action against us.

There was a question of what would happen, and this is what I think probably delayed me more than anything else. Or at least brought me to the conclusion not to take the hard option. I was concerned not so much about the effect on the Russians . . . but I was concerned by the effect on the Chinese. After all, we had fought the Chinese, in Korea, as a matter of fact; did you know one of Mao's sons was killed in Korea fighting Americans? He never mentioned it to me, but I knew that that had happened. Ah . . . and I was concerned about that. And my other reason for not taking the hard option at that point, the other major reason, I didn't know whether that kind of . . . of action . . . protective reaction after they had shot down the plane . . . taking out an airfield might escalate into a war. I figured having one war on our hands was enough without getting another war on our hands, particularly one in which the Chinese were very close and the Russians, too. But nevertheless there was an area where we disagreed.

Another area where Kissinger and I sometimes disagreed, we didn't disagree but where our temperaments were different, were with regard to what one should do after a tough decision is made. What should your attitude be? Ah . . . I'm a fatalist . . . basically . . . Kissinger is more, despite his enormous intellectual capability, is one who, ah . . . is . . . perhaps somewhat less fatalistic and more determinist in his views. But more emotional . . . interestingly enough . . . ah . . . although I too have emotions . . . I tend to hide them perhaps more than he does or to submerge them or to suppress them. Ah . . . but be that as it may . . . we won't try to psychoanalyze each other at the moment, ah . . . but Kissinger, I well remember, after we went into Cambodia, he was for going in, but here I made a decision on the spot, which we had not discussed before. He wondered about it at the time, but he totally supported it once it was made. We went to the Pentagon the day after the first movement, into two sanctuaries, and I asked the people at the Pentagon how many there were, and they said there were six, and I said, "Well, let's move into all of them," and I remember Westmoreland raised the point that he didn't know whether or not we could even handle two.

However, the other chiefs and the rest felt maybe we could handle all of them, particularly with the way that the South Vietnamese in their early days were fighting. So right then the decision was made to go into six, and it was one of the best decisions we made. Then came Kent State, which was a terrible emotional shock to me . . . ah . . . and a very great shock to Kissinger and of course a torment of abuse because the implication was that because we did Cambodia, three [Frost N.B. It was actually four.] students were killed in Kent State, that one followed the other, although the student body president of Kent State pointed out, when he came to see me at the White House, that while the Kent State tragedy partly was due to the disagreement about the war, that long before the war there were other issues that were stirring people up.

Nixon sometimes painted his colleague as a brilliant but precocious fellow who required the president's steadying hand to remain operational. Notice how, in an anecdote telling about what followed the tragic shootings at Kent State University, after the Cambodia incursion, Nixon "steadies" his excitable colleague.

NIXON: But I remember, right after that, Henry came in one day and said, "You know, I'm not sure that we should have gone into this Cambodia thing, and perhaps now has come the time when we should shorten the time and get out a little sooner." He wasn't seriously considering it, but he said, "I think"—and he always used to preface it by saying—"and I must warn you, Mr. President"—"that the situation that I hear from my colleagues from the colleges and the universities is very, very serious, and, ah . . . Cambodia is . . . it could have been a mistake." And I said, "Henry," . . . I said, "We've done it." I said, "Remember Lot's wife. Never look back." I don't know whether Henry had read the Old Testament or not. But I had, and he got the point. And from time to time Henry, who always supported an initiative once a decision was made and carried it out with a very firm and strong hand due to the fact that he was an intellectual, an honest intellectual, would always tend to try to reappraise a decision to . . . not to second-guess it so much as to say . . . well, I wonder whether we have made a mistake there so that we can avoid a mistake in the future. And then he would talk about it. He was tending to look back . . . I believe from history. When you can't do anything about it . . . then, I say, go forward after that.

Whenever he would come in and say, "Well, I'm not sure we should have done this or that or the other thing," I would say, "Henry, remember Lot's wife." And that would end the conversation.

FROST: In fact, you are different in the sense that he is, ah . . . probably more emotional, at least openly, publicly, than you. Ah, how many times, for instance, did he say he might resign?

NIXON: Oh, to me, he would hint it on occasion. How many times, ah . . . ah . . . not many, not often. Ah . . . when I say "he would hint it," the way it would come up would be, and this really cut him to the quick, he would get letters, for example, from members of the Harvard faculty, ah, or other Ivy League colleges, or people he respected—like most intellectuals, he only respected intellectuals. Ah, he couldn't abide fools. Ah, although he treated fools well at times. Sometimes better than he treated intellectuals who might be giving him a hard time. But be that as it may, in this case, he would come in and say, "Well, I, I just wonder if my usefulness isn't finished, I wonder if I shouldn't resign." Ah, ah . . . he would get . . . Henry, Henry was a man who, for example, in all of our conversations with foreign leaders, particularly at the summit, he was cold and controlled. He would use his temper rather than have his temper use him. And I was somewhat the same way. In fact, we were almost totally alike.

Notice how, in the following account, Nixon perhaps reveals more about himself than about Kissinger.

NIXON: Ah, on the other hand, he had to blow off steam from time to time. He had a tendency to, to get highly elated by some piece of good news and very depressed by something that he considered to be bad news. Ah, that doesn't mean that he was emotionally unstable, it simply means that having the kind of wide-ranging mind that he was a genius in this area, or intellectual has, and, ah . . . one of the . . . one of the characteristics of an individual with an exceptional mind is that he can see the heights and also see the depths. And he feels them both. And Henry was that way. Well, I, of course, don't contend that I'm a genius, ah . . . and so forth. I usually can see the heights, and could be, feel somewhat elated, although I try to restrain elation, because I always know that the, ah, as Churchill once said, that "the brightest moments are those that flash away the fastest." And so that when you're up today, you may be down tomorrow. I, that was my political experience. And, so . . . Henry would feel highly elated by a conversation that he'd had with Dobrynin, and then we'd have a bad development or negative development and he would be greatly depressed, and I'd say . . . "Well, Henry, the situation hasn't changed. We shouldn't have been as elated as we were yesterday, and we shouldn't be so discouraged today. Just keep plowing along."

For example, that long and tortuous process of negotiations with Vietnam. There was time after time that I was, I just felt for him so. I'd always put a little note, a handwrit ten note, whenever we could get it delivered by a courier to Paris, on the bottom, just to encourage him and to tell him I was thinking about him and the rest. Ah, and then the tortuous process, and this was even more difficult because by this time I had to spend . . . as he knew it, ah . . . time on Watergate, and I was getting a lot of heat from the press and from the Congress and the rest, and, ah, and ah . . . therefore, Henry had . . . didn't feel that he wanted to burden me as much with the day-to-day reports and activities, although I insisted he do so because I tend to compartmentalize things. I'll think about the problems of fighting the Congress, or on one area for a few hours, and then think about foreign policy for a few hours.

But, in this case on the Mideast, there was one occasion, particularly, when he felt that he should come home. And, ah . . . he sent back a message. He was terribly depressed. He just didn't think there was any way to break,

break the roadblock, which seemed to exist insofar as negotiations in this case between the Israelis and the Syrians. Both of them were being unreasonable, he thought. And ah . . . I had to send back and did send back a very strong, but . . . but . . . ah . . . also understanding letter, ah . . . cable in which I said that he must make one more try. Let me point out, if I'd been in his place, I'd probably have packed my bags and sent the message in the air that I was on the way home. Ah, Henry at least gave me some advance warning, and I was able to catch him before the plane took off. Ah, so he had reason to be discouraged, but, ah . . . many times I think that the way in the instant historians write about the Kissinger-Nixon relationship, they . . . they misread it to an extent because they, they take, for example, an emotional statement which, ah . . . he may not really mean. Like his, you call, how many times he's threatened to resign. He would come in and suggest that maybe he should because he was no longer useful. And, I, of course, then would say, "Look here, just stop all that talk, let's talk about the real thing."

FROST: That's why, that's why I phrased it the other way. How many times did he come in and say that maybe, maybe he should resign?

NIXON: Maybe a half a dozen.

FROST: Half a dozen times.

NIXON: That's right. But to others, more often. He would talk to Haig and to, ah, Haldeman about this. And this would be when he would be in fights with the bureaucracy. He couldn't stand the bureaucratic infighting. He had differences, you know, with Secretary Rogers. And, incidentally, this was a very painful thing for me because Rogers had been my friend. He was a personal friend. Henry, of course, was not a personal friend. We were, we were associates but not personal friends. Not enemies but not personal friends. Rogers was a personal friend. But, ah . . . Henry was fighting . . . first, they were . . . they were two very proud men. They were two very intelligent men. There could be only one person to handle some of these major issues, and where secrecy was involved, I mean, secret negotiations, it had to be Henry, ah, in the areas like Vietnam, China, Russia, and the Mideast. Now, in the case of Rogers, on the other hand, being a very proud man, ah, he did not resent Henry handling such things, but he objected to the fact that Henry got too much credit, and he felt was taking too much credit, and also he objected, and here I think he had a good case, and—I think Henry would have to agree—that he, Rogers, who had to make public statements all the time and testify and answer questions before the Congress, wasn't informed about things. He wanted to be informed.

Rogers may have wanted to be informed, but from his jealously guarded NSC perch, Kissinger thrived on the belief that knowledge is power. This usually meant that Rogers was kept ill informed about many a critical U.S. initiative, including in the Middle East and China.

NIXON: Well, Henry would come to me, and we had several arguments about it. He would say, "I will not inform Rogers, because he'll leak." And I say, "Henry," I must have told him this a dozen times. I said, "Henry, the State Department bureaucracy will leak. It always has. It always will." I said, "But Bill Rogers will never leak if I tell him it's in confidence." "Well, I'm not so sure." See, he didn't know Rogers as well as I did. I knew that Rogers was a man of honor, and I knew he wouldn't leak. And, that was why, on the China initiative, for example, we had a very good . . . we had an argument about that. Henry didn't want to tell anybody, of course, except those on a need-to-know basis. And I said, "Rogers has gotta know." And he said, "Well, he'll leak," or he said, "Well, he'll leak, or he'll object to it." I said, "You cannot have the secretary of state not be informed, because he has got to take off the day that announcement was made." And incidentally, that announcement took three and a half minutes when I announced that trip to China on July the fifteenth, 1971. But Rogers immediately had to call ambassadors all over . . . and heads of state all over the world, and Henry, of course, made a few calls and I made two or three.

I had heard that Kissinger had had the last laugh on Nixon, blocking the appointment of John Connally to replace Rogers as secretary of state, effectively ensuring that, to Nixon's chagrin, Kissinger would hold both positions in the second Nixon administration.

FROST: Did Henry say that he'd resign if John Connally was appointed secretary of state?

NIXON: Not to me. I, ah . . . I have read reports to that effect, and I do know that he, ah . . . that his views with regard to Connally were mixed. He had . . . he respected him as a political leader. However, ah . . . I think Henry, ah, saw in Connally, let's face it, a potential rival. Ah, Connally basically . . . everything that Connally touches, ah . . . in the political area, Connally controls. Henry's the same kind of a person. And so Connally would be a very formidable fellow to have around. Ah, and also you have to remember that Henry, to his credit, was loyal to his former patron and still his patron, Nelson Rockefeller. And he felt that building Connally up might, ah, not be in the interest of Nelson Rockefeller. But as far as his discussion with me, he never said to me, "Look, if ah . . . Rocke— . . . if Connally's appointed secretary of state, I'll resign" or "If you name Connally as the vice presidential nominee, I'll resign" or anything like that. Ah, I could sense that he would prefer somebody else, let's put it that way.

FROST: Were you actually considering John Connally as secretary of state?

NIXON: Yes, I thought he would have made a very good secretary of state. However, in this case, while Henry did not have a veto power, ah . . . nobody can have a veto power where the president is concerned, ah, any president. But while he didn't have a veto power, it was indispensable that whoever was secretary of state be able to work with Henry and Henry be able to work with him. Because he had his fingers in so many pies, ah . . . which were in various stages of development, and consequently, we couldn't possibly have a situation where he'd be at odds. In other words, I had gone through the Rogers-Kissinger feud of four years, and I didn't want to put another feud with another secretary of state for the . . . for the rest of the four years. And that's why I finally made Henry . . . gave Henry both hats, which I, ah . . . in retrospect, probably would not have done had, ah . . . we . . . could we have found some individual who would be Henry's equal or that would be considered his equal intellectually, and yet Henry would not feel was a competitor who would threaten his position of a . . . being the president's major foreign policy adviser. That position, he felt . . . incidentally, not just ego, but because frankly, he honestly felt he knew more about it than anybody else did. He honestly felt he was the best adviser. That didn't mean that from time to time, as we did at the time of the May 8 bombing, when, ah, we got advice from Connally and took his rather than Henry's, ah . . . with regard to go ahead with the bombing and don't cancel the summit, rather than the other way around, which is the way Henry first recommended it and the way I first approved it. But nevertheless, Henry felt that, ah . . . he had been . . . because of his experience, because of his background, and also because he was personally involved with so many leaders and they expected to deal with him, that he had to be the major foreign policy adviser, and he therefore couldn't tolerate a secretary of state who would impinge on that position.

Nixon and Kissinger worked miraculously well together and also got under each other's skin. Kissinger, in particular, was reported to have said things about Nixon that only a saint would have tolerated. As Nixon had few sainthood credentials, I could not wait to ask him about Kissinger's occasional indiscretions.

FROST: Now, you said you were not personal friends. You were not enemies, but you were not personal friends. Dustin Hoffman once made a film called *Who Is Harry Kellerman and Why Is He Saying Those Terrible Things About Me?* Now, as you read accounts, knowing what a successful working relationship you had in many areas, of remarks attributed to Henry Kissinger, whether it's in *The New Republic* or at the Ottawa Banquet, or whatever, you must sometimes feel, ah, you know who Henry Kissinger is, but did you sometimes feel, "Why is he saying all those terrible things about me?"

NIXON: Well, to answer the questions quite candidly, it drives my family right up the wall. And it's only because, that, it bothers them that it would bother me at all. After such accounts appear, I know that, ah . . . I always get a call from Henry on the phone, ah . . . ah . . . explaining that, ah . . . that there's been either a misquotation or misinterpretation or what have you. And I have always said to him, "Ah, forget it." I said, ah . . . ah . . . "What your opinion of me is, ah . . . is, ah . . . however you express it, it isn't going to affect our relationship unless you express it to me personally. I mean, what you say about me to other people isn't going to bother me."

That's what I said. But in all candor, I would have to admit my family didn't share that, and I think what we have to understand, too, is that Henry likes to say outrageous things. He's kind of like Alice Longworth that way. She, as she puts it, she said, "You know, I like to be naughty sometimes. I like to say things that are devilish." And, ah . . . ah . . . Henry was not really exactly that way, ah . . . in fact, it was quite different. Ah, but . . . he basically, as an intellectual, is, not typical in this sense. Ah . . . most people with great intellectual ability, ah . . . couldn't care less about the so-called Hollywood celebrity set or celebrities of any kind. I mean, basically their only interest is in a person's brains, and not particularly whether or not they have a lot of money or a lot of good news clippings or what have you. But Henry, on the other hand, was fascinated first by the celebrity set, and second, he liked being one himself. Not at first, but people would start coming up for his autograph and he was invited more and more to the Hollywood parties and the rest. And when you go to these parties, ah . . . I used to be in that category when I was vice president and even as a senator and congressman, a rather well-known one. And when the Georgetown hostesses had you in, they don't have you over in order to feed you . . . they have you there in order to . . . for you to entertain their guests with some little tidbit that has happened that day. "What Eisenhower said" or "what had Dulles said" or "What kind of a man is Churchill? Did you really talk to him?"

Well, anyway, that's Henry. Henry, when he goes to these parties, and he likes parties. I despise them because I've been to so many. I used to like them. But Henry will learn to despise them too after he's been through a few more. But be that as it may, so he goes to a party, and I can see exactly what happened in Canada. He runs into a lady who . . . has a very low opinion of me . . . and, ah, ah . . . so Henry feels that really he's defending me and that the way best to defend is to concede that, ah, "Well, he's sort of an odd person, he's an artificial person," and so forth and so on and so on. The only problem was that he didn't think to turn the microphone off. On the other hand, I didn't turn it off either in the Oval Office on occasions, so I never held him for that.